## CAMBRIDGE COUNTY GEOGRAPHIES
General Editor: F. H. H. GUILLEMARD, M.A., M.D.

# DEVONSHIRE

*Cambridge County Geographies*

# DEVONSHIRE

by

FRANCIS A. KNIGHT

AND

LOUIE M. (KNIGHT) DUTTON

With Maps, Diagrams and Illustrations

Cambridge:
at the University Press
1910

CAMBRIDGE UNIVERSITY PRESS
Cambridge, New York, Melbourne, Madrid, Cape Town,
Singapore, São Paulo, Delhi, Mexico City

Cambridge University Press
The Edinburgh Building, Cambridge CB2 8RU, UK

Published in the United States of America by Cambridge University Press, New York

www.cambridge.org
Information on this title: www.cambridge.org/9781107690752

First published 1910
First paperback edition 2013

*A catalogue record for this publication is available from the British Library*

ISBN 978-1-107-69075-2 Paperback

# PREFACE

I N preparing this book much use has been made of the *Proceedings* of the Devonshire Association and of the first volume of the *Victoria History of Devon*. The authors also desire to take this opportunity of recording their grateful thanks to Her Gracious Majesty Queen Alexandra for her kindness in providing one of the most interesting illustrations in the volume—the beautiful photograph of the Armada trophy preserved among the Royal plate in Windsor Castle, taken for the purpose of this volume by her command.

F. A. K. AND L. M. D.

*March*, 1910.

# CONTENTS

# ILLUSTRATIONS

## MAPS

The authors are indebted to Mr John S. Amery for leave to reproduce the pictures on pp. 154, 157 and 158.

# 1. County and Shire. The Name Devonshire.

The word "shire," which is probably derived, like "shear" and "share," from an Anglo-Saxon root meaning "to cut," was at one time used in a wider sense than it is at present, and was formerly applied to a division of a county or even of a town. Thus, there were once six small "shires" in Cornwall.

The word shire was in use at the time of King Ina, and occurs in the code of laws which that monarch drew up about the year 709; but the actual division of England into shires was a gradual process, and was not complete at the Norman Conquest. Lancashire, for example, was not constituted a shire until the twelfth century. Alterations in the extent and limits of some of the counties are, indeed, still being made; and in the case of Devonshire the boundaries have been changed several times within the memory of persons still living.

The object of thus dividing up the country was partly military and partly financial. Every shire was bound to provide a certain number of armed men to fight the king's battles, and was also bound to contribute a certain sum of money towards his income and the expenses of the state; and in each district a "shire-reeve"—or sheriff, as we call the officer now—was appointed by the Crown to see that the people did their duty in both respects. The shire was a Saxon institution. County is a Norman word, which came into use after the Conquest, when the government of each shire was entrusted to some powerful noble, often a count, a title which originally meant a companion of the King.

It has been suggested that the reason why the names of some counties end or may end in "shire," while in other cases this syllable is never used, is that the former were "shorn off" from some larger district, while the latter represent entire ancient kingdoms or tribal divisions. According to this theory, Yorkshire is a "shire" because it originally formed part of the kingdom of Northumbria; and Kent is not a "shire" because it practically represents the ancient kingdom of the Cantii. The form "Kent-shire" is, however, found in a record of the time of Athelstan.

In the case of our own county both forms are in use, and we say either "Devon" or "Devonshire," although the two names are not exactly interchangeable. Thus, while we generally talk of "Red Devon" cattle, we always speak of "Devonshire" cream. "Devon," which is the older form, may be derived either from *Dumnonii*, the name given by Ptolemy, an Alexandrian geographer of the second century, to the inhabitants of the south-west of Britain, perhaps from a Celtic word *Dumnos*, "people"

—or it may come from the old Welsh word *Dyvnaint* or *Dyfneint*, "the land of the deeps," that is to say, of deep valleys or deep seas. To the Saxon settlers the people they found in possession of the district were *Defn-saetan* or "dwellers in Devon"; and in time these settlers called themselves *Defenas*, or "men of Devon." In the Exeter Domesday Book—the Norman survey of the five south-western counties, completed probably before 1086—the

**Devonshire in the Exeter Domesday Book**

name of the county is given as *Devenesira*. It would appear, then, that the Britons called their province "Devon," and that the Saxons called it "Devonshire." It is characteristic of the peaceable nature of the Saxon occupation that the two names, like the two nations, seem to have quietly settled down side by side.

It is believed that it was Alfred the Great who marked out the border-line between Devon and Somerset; and it was undoubtedly Athelstan who, after his victory over the West Welsh, made the Tamar the boundary between Devon and Cornwall.

## 2. General Characteristics.

Devonshire is a county in the extreme south-west of England, occupying the greater part of the peninsula between the English and Bristol Channels, and having a coast-line both on the south and on the north. Situated thus, on two seas, and possessing, especially on its southern sea-board, a remarkable number of bays and estuaries, it has always been noted as a maritime county. And although many of its harbours have, in the lapse of ages, become silted up with sand or shingle, and are now of comparatively slight importance, it has one great sea-port, which, while only thirtieth in rank among British commercial ports, is the greatest naval station in the Empire.

The county has in the past been famous for its cloth-weaving and for its tin and copper-mining, but these industries are now greatly decayed, and the main occupation of the people is agriculture, to which both the soil and the climate are particularly favourable.

A special characteristic of Devonshire is its scenery, which is so striking that it is very generally considered the most beautiful county in England; while there are probably very many who regard its mild and genial, equable and health-giving climate as more noteworthy still. It is a remarkably hilly country, and it also possesses not only many rivers, but a great number of broad river estuaries. Another characteristic with which every visitor to the district is struck is the redness which distinguishes

King Tor, near Tavistock

its soil, its southern cliffs, and its famous breed of cattle, which is not less noticeable than the soft and pleasant dialect, with its close sound of the letter "u" so typical both of Devon and of West Somerset.

Another characteristic of the people has always been their loyalty to their sovereign, to their county, and to

A Typical Devon Stream—Watersmeet, Lynmouth

each other. Devon is proverbial, like Cornwall and Yorkshire, for the clannishness of its inhabitants. It is a land, too, where superstition dies hard. Belief in pixies—fairies, as they are called elsewhere—in witches and witchcraft, in whisht-hounds and other weird and uncanny creatures, and in portents and omens, still lingers, especially on Dartmoor.

Dartmoor itself, with its wild and picturesque scenery, its unrivalled wealth of prehistoric antiquities, and its singular geological structure, forms one of the most striking features of the county, and one to which there is no parallel in England. The marine zoology of Devonshire is more interesting than that of any other

A Devon Valley—Yawl Bottom, Uplyme

English county, and nowhere else in the island has there been discovered clearer evidence of the great antiquity of man than was found in Kent's Cavern and other Devonshire caves.

Above all things, its position has made Devonshire a native land of heroes. Very few other counties have produced so many men of mark, so many men of enterprise

and daring.   Certainly no other has played a greater part
in the expansion of England.   From Devonshire came not
only some of the most distinguished seamen of the Golden
Age of Elizabeth, some of the most skilful and daring
of her naval captains, but some of the earliest and most
famous of our explorers; the founder of the first English
colony, the first Englishman to sail the polar sea, the first
Englishman to circumnavigate the globe.

## 3.   Size.   Shape.   Boundaries.

Devonshire, which occupies rather more than one-
twenty-second of the whole area of England and Wales,
is one of the largest counties in the British Islands, being
exceeded in size only by Yorkshire and Lincoln in England,
by Inverness and Argyll in Scotland, and by Cork in
Ireland.   Its extreme length from east to west, measured
along a horizontal line drawn through the middle of the
county, starting at the Dorsetshire border half-way
between Lyme Cobb and the Seven Rocks Point, passing
close to the city of Exeter, and reaching to the point
where the river Ottery enters the county, is 67 miles;
exactly the same as that of the county of Somerset.   Its
greatest breadth, from Countisbury Foreland on the north
coast to Prawle Point on the south, is 71 miles.   It may
be added that a longer east and west line can be drawn
only in Yorkshire and Sussex, and a longer meridional
line only in Yorkshire and Lincoln.   The area of the
"Ancient" or "Geographical" county of Devonshire,

according to the revised return furnished by the Ordnance Department, is 1,667,154 acres, or 2605 square miles. Compared with the counties that adjoin it, it is two-and-a-half times the size of Dorset, it is roughly twice as large as Cornwall, and it is more than half as large again as Somerset.    It is fifteen times as large as Rutland, it is

Glen Lyn, near Lynmouth

about half the size of Yorkshire, and its area is less than that of Lincolnshire by only 48 square miles.

Although usually said to be irregular in form, the outline of the county has a certain degree of symmetry, being roughly shaped like a life-guardsman's cuirass, with nearly equal sides, with a small hollow at the top or north coast, and a much larger one at the bottom or south coast.

Devonshire, like Kent and Cornwall, is bounded on two sides by the sea, having the Bristol Channel on the north and north-west, and the English Channel on the south. On its western side the river Tamar, with its tributary the Ottery, forms almost the whole of the frontier between it and Cornwall. The eastern and north-eastern border is less definite, but is roughly marked by Exmoor and the Blackdown Hills, which partly separate Devonshire from Somerset. The short length of frontier between Devonshire and Dorsetshire is marked by no natural feature.

No part of Devonshire is now, as was formerly the case, wholly surrounded by any other county. Three of its parishes, however, are partly in Dorset, one is partly in Cornwall, and one, a district of Exmoor containing no houses or inhabitants, is partly in Somerset. Culmstock, which before 1842 was considered to belong to Somerset, although completely islanded in Devon, and Stockland and Dalwood, which were reckoned with Dorset, although they were entirely inside the Devonshire border, have now been formally transferred to this county. On the other hand, Thornecombe and Ford Abbey, which belonged to Devonshire although they were situate in the adjoining county, have been handed over to Dorset. Still later alterations were the transfer of Hawkchurch and Churchstanton from Dorset to Devon in 1896.

## 4.  Surface and General Features.

Devonshire is characterised by such great irregularity and unevenness of surface that practically the only level land in it is along the shores of its estuaries; with the almost inevitable result that it is one of the most picturesque and beautiful counties in England.   Its scenery has been very greatly affected by subterranean movements, which have not only roughly shaped its hills and valleys, partly by upheaval and partly by the shrinkage of the earth's crust, but have been the principal cause of the breadth of the river estuaries which are so marked a feature of its coasts, especially of the south.   At many points along the shore of Devonshire there is evidence, in raised sea-beaches, and, near Torquay, in the borings of marine mollusca at a great height above the present tide-line, of upheavals that must have raised the whole coast, even if they did not materially change the contour of the country.   On the other hand, the existence of submerged forests at many places near the shore proves that the land has sunk at least forty feet, thus allowing the sea to flow further inland; thereby greatly widening the already existing valleys, which had been formed in part by the shrinkage of the earth's crust, and in part by the action of the rivers.

The chief physical feature of Devonshire, a feature without parallel in any other part of England, is the Forest of Dartmoor, the great upland, some twenty miles long and eighteen miles broad, which occupies so large a

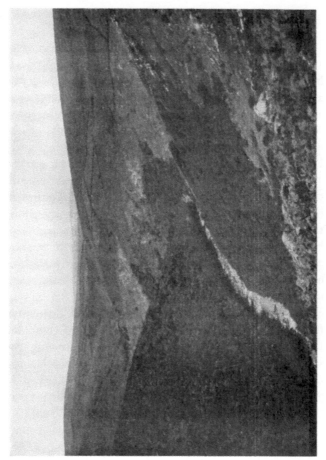

The Upper Dart, from the Moors

part of the southern half of the county.  It is all granite, the largest mass of granite in England, and forms part of a chain of outcrops of that formation extending from Devonshire to the Scilly Isles.  The word "forest," it should be remembered, originally meant, not a wood, but a hunting-ground.  No part of the open moor is now covered with trees, nor is it likely, considering the poorness of the soil, that it ever was so covered, although roots and other remains of trees have been found in various parts of it.  In early days it was a royal hunting-ground, and most of it is still Crown property, forming part of the Duchy of Cornwall.

The most prominent feature of the moor, which contains the highest ground in England south of Ingleborough in Yorkshire, are the isolated rocky heights called tors, some 170 in number, many of which have been weathered, not only into very rugged and highly picturesque, but even into most strange and fantastic shapes; in many cases having their steep slopes strewn with fallen fragments of rock, some of them tons in weight, forming what are known on the moor as "clitters" or "clatters."  The highest points are High Willhays, 2039 feet; Yes Tor, 2029 feet, only half a mile away from its rival, Newlake, 1983 feet; Cuthill, 1980 feet; and Great Lynx Tor, 1908 feet above sealevel; and among the most striking and picturesque are Great Lynx Tor, Staple Tor, Mis Tor, and Vixen Tor, although many others are remarkable for their strange and time-worn outlines.

The moor is seamed by many valleys and ravines,

not a few of which are, in parts, well-wooded, each with its swiftly-flowing stream or river, and many of them most picturesque and beautiful. Such, in particular, are the Valley of the Dart, especially including Holne Chase and above; of the Teign near Fingle Bridge; of the Tavy at Tavy Cleave; of the Lyd at Lydford, and of the Plym at the Dewerstone.

Tavy Cleave, showing disintegrated granite

Dartmoor is distinguished in being the coldest and rainiest part of Devonshire, and to these two features of its climate are no doubt largely due the fogs which so frequently envelope it. Its great extent and its heavy rainfall make the moor the main watershed of the county. Most of its rivers have their sources in the bogs, which are a well-known and somewhat dangerous feature of the district, and of which the most remarkable are Fox Tor Mire, Cranmere Bog, and Cuthill Bog.

Its varied and peculiar features, its vast expanses of wild and desolate moorland, now aglow with golden gorse, and now still more splendid with the magnificent purple of its broad sheets of heather or with the warm hues of dying bracken, and beautiful, as the seasons change, with the varying tints of grass and sedge, of ferns and rushes, of moss and bog-myrtle and bilberry, of cotton-grass and asphodel; the almost unrivalled beauty of its river-valleys, its multitudinous streams, its wild life, its extraordinary wealth of prehistoric antiquities, its lingering superstitions of pixies, of witch-craft, of night-flying whisht-hounds and ghostly huntsmen, its very solitude and silence, combine to make Dartmoor, to the antiquary and the artist, the naturalist and the angler, one of the most attractive spots in England, and one whose charm poets, painters, and authors have striven from earliest days to immortalise.

The greater part of Exmoor, and all its principal heights, are in Somerset, but it extends into the north-eastern corner of Devon, and detached portions of it, which appear to be really parts of the same upland, reach to the hills above Combe Martin. Part of Span Head, whose summit is 1619 feet above the sea, is in our county; and the outlying spurs of Bratton Down, Kentisbury Down, and the Great Hangman are all over 1000 feet high. There is very beautiful scenery on Exmoor, especially on the Somerset side of the border, somewhat resembling that on Dartmoor, although less

wild and picturesque, and without any of the tors which
are so characteristic of the greater upland. Exmoor is
the only part of England where red deer still run wild ;
and the district is visited every year by stag-hunters from
all parts of the island and especially from Ireland. Both
it and Dartmoor are famous for a breed of sturdy little
ponies, originally, no doubt, of the same stock. In the
Badgeworthy Valley, which is in Somerset, although not
far from Lynton, may be seen what are said to be the
ruined huts of the Doones, a community of freebooters
immortalised by Blackmore, who represents them as
having been the terror of the country-side towards the
close of the seventeenth century.

Other Devonshire hills are the Black Downs, along
the border of Somerset, in which the highest point is
860 feet above the sea; another Black Down, six miles
due south, reaching 930 feet; the Great Haldons, south-
west of Exeter, 817 feet high; and Dumpdon Hill, about
two miles north by east of Honiton, 856 feet above sea-
level.

Devonshire is in parts extremely fertile, especially
towards the south, and it has been called (in common, it
is true, with other counties) the Garden of England. Two
very large and specially productive areas are the Vale of
Exeter, and the South Hams,—the latter a name somewhat indefinitely applied to the district south of Dartmoor
and occupying a large part of the region between the
Teign and the Plym, with Kingsbridge as its chief centre.
The great fertility of this famous district is due partly
to the nature of the soil, partly to the mildness of the

On Lundy

climate and the shelter afforded by the heights of Dartmoor, and partly to its nearness to the sea.

A very remarkable and interesting feature of Devonshire is Lundy—an island three miles long by one mile broad, lying out in the Bristol Channel, opposite Barnstaple Bay, and twelve miles north-north-west of Hartland Point. Its name, it is believed, is derived from two Norse words meaning Puffin Isle.

Composed entirely of granite, except for its southern extremity, which is millstone grit, its lofty cliffs are very wild and rugged and picturesque, and for two miles along its eastern side there is a remarkable series of chasms, from three to twenty feet wide and some of them of great depth, known to the islanders as the Earthquakes. The shingle beach at the south-eastern corner, in the shelter of Rat Island, is the only landing-place, but many vessels find good anchorage on the eastern side, well protected from westerly winds. Many ships, however, have been wrecked among the terrible rocks round its base, including the battleship *Montagu*, lost in 1906, and, according to tradition, one of the galleons of the Spanish Armada. There is a lighthouse at each end of the island, and the southern one is the most powerful in Devonshire.

Perhaps the greatest charm of Lundy lies, as will be shown in some detail in a later chapter, in its natural history, especially in the vast numbers of birds which visit it in the breeding season. Among very rare stragglers that have been shot here is the Iceland falcon, a species of which very few examples have been recorded for this country. A few plants and insects are peculiar to the spot. There are now few trees, except those planted not long ago near the owner's house in a cleft at the south-eastern end, but some shrubs, such as fuchsias, hydrangeas, and rhododendrons grow to a great size, and the mesembryanthemums are particularly vigorous and beautiful.

Granite for the Thames Embankment was obtained here, but the quarries have long been closed, and farming is the chief industry of the few inhabitants.

There are evidences of very ancient occupation, in the shape of kistvaens, tumuli, and the foundations of primitive dwellings; and in times more recent the island has had a stirring history. In the reign of Henry II it was held by the turbulent family of the Montmorencies or Moriscos, and the shell of Morisco Castle, now converted into cottages, still stands on the south-east corner of the island. During the Civil War it was fortified for the king, and only surrendered in 1647. At various times in the seventeenth century it was captured by French, Spaniards, and Algerines; and it was, moreover, several times occupied by pirates, some of whom were Englishmen, who found it a convenient station from which to plunder ships sailing up the Bristol Channel.

## 5. Watershed. Rivers and the tracing of their courses. Lakes.

Devonshire is a well-watered county, a county of many rivers; and although not one of its multitudinous streams is of real commercial importance or of much value as a water-way, by their mere abundance and by the beauty of their scenery, especially of the magnificent ravines which many of them in the lapse of ages have worn deep in the rock, they form one of its most striking features.

By far the most important watershed is the great upland of Dartmoor, where, with few exceptions, rise all the principal rivers. The headwaters of the Tamar and the Torridge—which rise close together, but flow in very different directions and reach different seas—are in the high ground in the north-west, on the very border of Cornwall, and the sources of the Exe and of its great twin stream the Barle are on the moor to which the former gives its name, just inside the county of Somerset. But the tributaries of all these are drawn from the bogs of Dartmoor, and especially from the morasses round the now insignificant sheet of water known as Cranmere Pool. The whole eastern border of the county, from Exmoor southward to the Blackdown Hills, is a source of streams. Such are the Lyn, flowing into the Bristol Channel; the Bray, the Yeo, and the Mole, tributaries of the Taw; the Loman, the Culm, and the Clyst, tributaries of the Exe; the Otter, falling into the English Channel; and the Yarty, a tributary of the Axe. It is remarkable that of all the many streams of Devonshire, only two of any consequence reach the estuary of the Severn.  Almost all flow into the English Channel.

The longest of the Devonshire rivers is the Exe, after which are named Exford and Exton in Somerset, and Exeter and Exmouth in our own county—a strong and beautiful stream which rises near Simonsbath on Exmoor,

The River Exe at Tiverton

flowing for the first twenty miles through Somerset and crossing the Devonshire border near Dulverton station, where it is met, on the left bank, by its great tributary the Barle.  It then runs nearly due south, through well-wooded and fertile country, being joined on its left bank, at Tiverton, "the town of the two fords," by the Loman; and farther down on the same side by the Culm, which gives its name to Culmstock.  Near Exeter it receives on

the right bank the Creedy, a pretty and winding stream
that lends its name to Crediton, and along whose shores
is some of the richest land in Devonshire. A little below
Exeter, close to the once famous port of Topsham, it is
joined on the left bank by the Clyst, a small and unimportant stream, flowing through most fertile country, and
giving its name to no fewer than seven villages. Below
Topsham the Exe widens out to nearly a mile, forming,
at high tide, from this point to the sea, a noble estuary
five miles long, with the popular watering-place of
Exmouth on the slope of the eastern side of its entrance,
which is almost closed by a long sandbank called the
Warren, divided into two parts by a stream. Until late
in the thirteenth century the Exe was navigable from
the sea to Exeter. But in 1290 Isabella de Fortibus,
Countess of Devon, having quarrelled with the citizens,
blocked the river-bed with stones, at a place still called
the Countess Weir, leaving, however, sufficient room for
ships to pass. At a later period this space was closed by
the Earl of Devon, and the navigation of the river entirely
stopped. Vessels now reach Exeter by a canal.

The second river in point of length is the Tamar,
after which are named North Tamerton in Cornwall and
Tamerton Foliott in our own county. Rising in the
extreme north-west, in the high ground that parts
Devonshire from Cornwall, it forms almost the whole
of the dividing line between the two counties, and is
characterised throughout the lower portion of its course
by some very beautiful scenery. It is joined by many
streams, some rising in Devonshire and some in Corn-

wall; some of which—the Lyd, for example—are renowned for their wildness and beauty. The largest of the western tributaries is the Lynher, entirely a Cornish river, whose estuary joins the Hamoaze. The most important of those on the left bank is the Tavy, a Dartmoor-drawn stream, giving its name to the town of Tavistock and to the villages of Peter Tavy and Mary Tavy, and flowing through some of the most fruitful land in Devonshire. A particularly fertile district is that lying between the Tavy and the Tamar.

Although it is a much shorter river than the Exe or the Tamar, the Dart is better known than either, and is perhaps the most familiar by name of all the Devonshire streams. Along its banks, especially near Holne and Buckland-in-the-Moor, and along the wooded shores of its magnificent estuary, is some of the most beautiful river-scenery, not in this county only but in all England. The most important of its many tributaries are the East and the West Dart—both of which rise in the great bog round Cranmere Pool, and join at a picturesque spot called Dartmeet—and the Webburns, East and West. Below Totnes the Dart widens out into a long and most beautiful estuary, winding among finely-wooded hills. On the west side of its entrance is the old port of Dartmouth, named, like Dartington, after the river, and on the opposite shore is the smaller but equally picturesque little town of Kingswear.

Famous as the Dart is for the wildness and beauty of its scenery, and for the excellence of its trout and salmon fishing, it has an evil name for the dangerous nature of

On the Dart; Sharpham Woods

its swiftly-flowing waters, which, after heavy rain on the moor, rise with extraordinary rapidity, changing it in a few hours from a peaceful and easily-forded stream into a raging and resistless torrent.  At Hexworthy, in November, 1894, the river rose ten and a half feet above the level of the previous day.  Characteristic of this as of other of the moorland streams, is the strange sound it sometimes makes, especially towards nightfall, known as its " cry," and believed by the superstitious to be ominous of flood and danger.  To " hear the Broadstones crying " —masses of granite lying in the bed of the stream—is considered by the moor-folk a sure sign of coming rain.

The Dartmoor rivers, in the upper part of their courses, are naturally all swift, and are all more or less tinged by the peat of their moorland birthplace—lightly, when the stream is low, and deepening in flood-time into the colour of a rich cairngorm.

The Teign, another of the streams that rise in the Cranmere bog, is famous both for the beauty of the scenery along its winding shores and for the many prehistoric antiquities—stone circles and alignments, menhirs and tumuli—which stand near them.  Its two main branches, the North and the South Teign, meet about a mile to the west of Chagford.  To the east of that moorland village the river flows through beautifully wooded valleys, and is joined on its right bank, below Chudleigh, by another Dartmoor tributary, the Bovey, on which stand Bovey Tracy, famous for its beds of lignite and clay and for its potteries, and North Bovey, near which are the remains of the very remarkable Bronze

Age village of Grimspound. Below Newton Abbot the Teign becomes a broad estuary, on or near whose shores are five of the townships that are named after the river, the most important of which is the little port and well-known watering-place of Teignmouth. The river mouth is almost blocked by a low promontory, which, although

The Axe at Axminster Bridge

now built over, was once a mere sand-bank or dune, from which latter word, no doubt, it takes its name of the Den.

Other south-coast rivers are the Axe—one of whose two main branches rises in Somerset and the other in Dorset—which gives its name to Axminster and Axmouth; the Otter, which rises in the Blackdown Hills, and flowing past Honiton, Ottery St Mary, and Otterton, reaches the sea at Budleigh Salterton; the Aune or Avon,

especially famous for its salmon, the Erme, and the
Yealm, small but beautiful streams rising on the southern
slopes of Dartmoor, widening into estuaries as they near
the English Channel, and giving names to Aveton,
Ermington, and Yealmpton, respectively. The Plym,
after which are named Plympton, Plymouth, and Plymstock, is another Dartmoor river, flowing through some
very beautiful country, especially in the neighbourhood
of Bickleigh, and at length forming a broad and important
estuary known first as the Laira, and lower down as the
Catwater or Cattewater, which joins Plymouth Sound.

The chief rivers on the north coast are the Torridge
and the Taw, the former of which, rising in the extreme
north-west, on the Cornish border, near the source of the
Tamar, flows south-west for nearly half its course, and
then sweeps round to run in the opposite direction, giving
its name to three several Torringtons, and having as its
chief tributaries the Walden, the Lew, and the Okement,
all on its right bank. The last-named stream is formed
of the East and the West Okements, which meet at
Okehampton, their namesake. The lower waters of the
Torridge form a long and narrow estuary—its shore only
ten miles distant from the original source of the river—
half-way down which is the once important port of
Bideford, built on both sides of the stream, which is
here spanned by a very ancient bridge. Near the entrance
of the estuary, but neither of them on the open sea, are
Appledore, the port of Barnstaple, and Instow, a small but
growing watering-place.

The Taw is a Dartmoor-drawn river, rising, like so

many streams, in the Cranmere bog, giving its name to Tawstock and to three several Tawtons, and receiving on its right bank the Yeo, the Little Dart, and the Mole. The most considerable town on it is Barnstaple, beyond which it becomes a broad tidal estuary, joining that of the Torridge, and flowing out into what is known both as Barnstaple and Bideford Bay.

Bideford and the Torridge Estuary

Many small streams fall into the Bristol Channel, among which is the Lyn, renowned for its beautiful scenery and its good trout-fishing.

A large proportion of the Celtic words in our language are found in the names of natural features, especially of hills and rivers. This is particularly well seen in Devonshire, where, as has been pointed out, the Saxons came as settlers rather than conquerors, adopting many of the names which they found already in use, and where an unusually large number of towns and villages have been called after the streams on which they stand.

The names Exe, Axe, and Okement, from the Celtic *uisge*; Avon, Aune, and Auney, from *afon*; Dart, from *dwr*; and Teign, from *tain*, are all derived from roots meaning "water." Other names are taken from descriptive adjectives, such as Wrey, from *rea*, rapid; Lyn, from *lleven*, smooth; and Tamar, Taw, and Tavy, from *tam*, spreading or still.

The lakes of Devonshire, as is the case in the majority of English counties, are little more than ponds. Cranmere Pool, in the great morass where many Devonshire rivers rise, lying in a dreary spot, as befits the reputed place of punishment of evil spirits, has shrunk of late years in consequence of much peat-cutting in its neighbourhood, and is now an insignificant pond, rarely more than seventy yards across, and in hot summers sometimes quite dry. Bradmere Pool and Classenwell Pool, the sites of old mine-workings, are beautiful little lakes, but they are only a few acres in extent. Burrator Reservoir has been made in order to supply water to Plymouth. The largest of these miniature lakes is Slapton Ley, or Lea, a long and narrow sheet of water, two and a quarter miles in length and measuring about 200 acres, separated from the sea, with which it was no doubt once connected, by a bank of fine shingle. The reeds of its north-eastern end, which are cut and sold for thatching, are the haunt of many water-birds; and the Ley is visited in winter by immense numbers of migratory ducks and waders.

## 6. Geology.

Three main points characterise the geological features of Devonshire; the simplicity of the system in the west, north-centre and south-west of the county; the comparative complexity and variety of the strata in the east and south; and, most remarkable of all, the extraordinary number of outcrops of igneous rock, from the great mass of Dartmoor granite, which has no parallel in England, to the hundreds of small dykes or elvans that are scattered chiefly over the southern region, although some occur to the north and east of Dartmoor.

The oldest rocks in Devonshire are probably not, as was once thought, the granites, but the highly altered or metamorphic formations in the extreme south; that is to say, the mica and quartz schists and the hornblende epidote schists which extend from near Start Point to Bolt Tail, a district which, owing in great measure to distortion by volcanic upheaval, includes some of the most picturesque scenery in Devon.

Next in order of age is the series called Devonian, after the name of the county, in which they were first distinguished from the Old Red Sandstone. They are, however, by no means confined to Devonshire, but are very widely distributed, covering a large part of Cornwall, and occurring on the continent of Europe, especially in Russia, and in Asia and North and South America. The Devonian beds—which are found both in the north and south, occupying two distinct areas separated by wide-

| | NAMES OF SYSTEMS | SUBDIVISIONS | CHARACTERS OF ROCKS |
|---|---|---|---|
| **TERTIARY** | Recent<br>Pleistocene | Metal Age Deposits<br>Neolithic ,,<br>Palaeolithic ,,<br>Glacial ,, | Superficial Deposits |
| | Pliocene | Cromer Series<br>Weybourne Crag<br>Chillesford and Norwich Crags<br>Red and Walton Crags<br>Coralline Crag | Sands chiefly |
| | Miocene | Absent from ·Britain | |
| | Eocene | Fluviomarine Beds of Hampshire<br>Bagshot Beds<br>London Clay<br>Oldhaven Beds, Woolwich and Reading<br>Thanet Sands [Groups | Clays and Sands chiefly |
| **SECONDARY** | Cretaceous | Chalk<br>Upper Greensand and Gault<br>Lower Greensand<br>Weald Clay<br>Hastings Sands | Chalk at top<br>Sandstones, Mud and<br>Clays below |
| | Jurassic | Purbeck Beds<br>Portland Beds<br>Kimmeridge Clay<br>Corallian Beds<br>Oxford Clay and Kellaways Rock<br>Cornbrash<br>Forest Marble<br>Great Oolite with Stonesfield Slate<br>Inferior Oolite<br>Lias—Upper, Middle, and Lower | Shales, Sandstones and<br>Oolitic Limestones |
| | Triassic | Rhaetic<br>Keuper Marls<br>Keuper Sandstone<br>Upper Bunter Sandstone<br>Bunter Pebble Beds<br>Lower Bunter Sandstone | Red Sandstones and<br>Marls, Gypsum and Salt |
| **PRIMARY** | Permian | Magnesian Limestone and Sandstone<br>Marl Slate<br>Lower Permian Sandstone | Red Sandstones and<br>Magnesian Limestone |
| | Carboniferous | Coal Measures<br>Millstone Grit<br>Mountain Limestone<br>Basal Carboniferous Rocks | Sandstones, Shales and<br>Coals at top<br>Sandstones in middle<br>Limestone and Shales below |
| | Devonian | Upper }<br>Mid } Devonian and Old Red Sand-<br>Lower } stone | Red Sandstones,<br>Shales, Slates and Lime-<br>stones |
| | Silurian | Ludlow Beds<br>Wenlock Beds<br>Llandovery Beds | Sandstones, Shales and<br>Thin Limestones |
| | Ordovician | Caradoc Beds<br>Llandeilo Beds<br>Arenig Beds | Shales, Slates,<br>Sandstones and<br>Thin Limestones |
| | Cambrian | Tremadoc Slates<br>Lingula Flags<br>Menevian Beds<br>Harlech Grits and Llanberis Slates | Slates and<br>Sandstones |
| | Pre-Cambrian | No definite classification yet made | Sandstones,<br>Slates and<br>Volcanic Rocks |

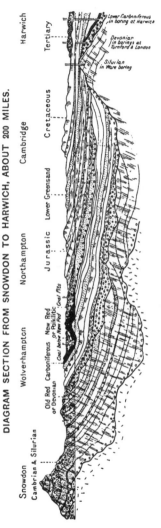

## DIAGRAM SECTION FROM SNOWDON TO HARWICH, ABOUT 200 MILES.

Snowdon
Cambrian & Silurian

Wolverhampton

Old Red Carboniferous New Red
or Devonian or Palkitic
Coal below New Red

Coal Pits

Northampton

Jurassic

Lower Greensand

Cambridge

Cretaceous

Harwich

Tertiary

Lower Carboniferous
in boring at Harwich

Devonian
in borings at
Turnford & London

Silurian
in Ware boring

This cross section shows what would be seen in a deep cutting nearly E. and W. across England and Wales. It shows also how, in consequence of the folding of the strata and the cutting off of the uplifted parts, old rocks which should be tens of thousands of feet down are found in borings in East Anglia only 1000 feet or so below the surface.

spread deposits of culm or carboniferous measures—were, it is thought, formed in open water, and probably at the same time that the Old Red Sandstone of the adjoining county of Somerset and elsewhere, which is not found in this county at all, was being deposited in estuaries and land-locked seas.

The North Devonian beds, which extend from the coast as far south as the latitude of Barnstaple, consist of slates, grits, and sandstones which, it is believed, judging from the organic remains in them, were formed in shallow water and near shore. Their lower strata, the Foreland grits, Lynton beds, and Hangman grits, contain some fossils and various kinds of coral. But the Middle beds, the Ilfracombe and Morte slates, are much richer in animal remains; of which perhaps the most remarkable are primitive palaeozoic fish, such as the very curious armoured *pteraspis*; while corals and bivalve shells are abundant and characteristic. The Upper Devonian is less fossiliferous, but contains some large trilobites, various marine shells, and some land-plants.

The South Devonian, which covers nearly all South Devon and a large part of Cornwall, is somewhat different in character, consisting chiefly of slates, with coralline limestones, varied by volcanic outcrops or elvans—a word said to be of Cornish origin, and meaning "white rock." To judge from its fossils, it was deposited in deeper water than the contemporary beds in the north of the county. The Lower and Middle beds are also far richer in animal remains; and the Middle Devonian of the south, which is the most typical of the series and includes the limestones of Plymouth and Torbay, is crowded with shells, trilobites, and corals. Among the shells, bivalves—such as *Stringocephalus*, which occurs only in the Devonian formations—spiral univalves, and corals are very abundant. There are also many crinoids, distinct from those of the carboniferous limestone, while perhaps the most charac-

Logan Stone, Dartmoor

teristic form is the rare and curious *Caleola sandalina*, differing from all other corals in having an operculum. There are not many varieties of trilobite, but the large *Brontes flabellifer* is not uncommon.

The Lower beds of this series contain fewer organic remains, although a good many fossils are found, including fragmentary remains of various fishes which have not yet been identified. The Upper Devonian is, on the whole, very poor in fossils.

Between the two Devonian areas, and occupying a large part of the centre of the county, are the carboniferous or coal-bearing measures, containing, however, not true coal but anthracite, which has more carbon in it than is found in ordinary coal; and these beds are perhaps more often known as Culm, from the Welsh *cwlwm*, a knot, in allusion to the fragmentary condition in which the mineral is frequently found. Anthracite, which elsewhere and especially in South Wales is a most valuable fuel, is here clayey and impure, and in thin seams. It is worked to a small extent, to be ground and made into a paint called Bideford Black. The Culm measures consist of grits, shales, and sandstones, with beds of chert and limestone containing fossil plants and other forms of marine life. Fish are few, only two species having been identified. The anthracite occurs in the middle Culm, and there are other remains of plants in both the middle and upper beds. The upper Culm is well seen on the coast near Clovelly and by the river Torridge, where it has been bent by volcanic upheaval into curious and beautiful curves. These measures, in general, are characterised by many outcrops of volcanic rock, some of which were probably contemporary, that is to say, they were poured out while the culm was in process of formation; while others are intrusive, or were forced up through the strata after these had been solidified into rock. These igneous rocks are found in great variety.

By far the most important and striking of these

volcanic formations is the great granite mass of Dart-moor, one of the most prominent features of the county, measuring 225 square miles in extent, and constituting the largest granitic area in England. Granite is a volcanic rock, formed, it has been suggested, by fusion at a great depth and under great pressure, and consisting in the main of three minerals, quartz, felspar, and mica. That

A smoothly-weathered granite Tor, Dartmoor

of Dartmoor is, on the whole, grey and coarse-grained, but it varies a good deal in colour, fineness, and composition. Its real origin is obscure. It has been assigned by various experts to various periods, and it has been called "the sphinx of Devon geology." There can, however, be no doubt about the great disturbance which has been caused in the county by upheaval and by the intrusion of melted rock, which has bent, broken, and

twisted previously-existing formations in a most extraordinary manner, the results of which are well seen in
the picturesque scenery of the Start, Prawle Point, and
Bolt Head. Lundy, which is twelve miles from the nearest
point of Devonshire mainland, is all granite, except for
a small part of its south end, which is Millstone Grit.

A long interval of time appears to have followed the
laying down of the Culm measures, during which so vast
an amount of shattered rock was worn away that when
the beds that come next in order—the New Red Sandstones—were formed, they were, in places, deposited
directly upon the Devonian, the superincumbent carboniferous or Culm strata having entirely disappeared. The
New Red Sandstones occur chiefly in the east of the
county, where their lower beds fill up old creeks and
valleys in the carboniferous system; and they extend
northwards from the coast past Exeter as far as Holcombe
Regis, forming broad bands on either side of the Exe,
characterised by the high fertility of the overlying soil,
and with one long spur traversing the heart of the county,
past Crediton and Exbourne, with isolated patches round
Hatherleigh, and with another and less extended pro-
longation a few miles west of Tiverton. The Lower
New Red consists of clays, conglomerates, red breccias
and sands, in which occur many outcrops of trap, the
evidence, not only of numerous eruptions, but of eruptions
extending over a long period of time. These beds contain no fossils, except in fragments of older rocks. The
Middle New Red, in the form of thick beds of red marl
and red and white limestones, well seen on the south

coast, is covered in turn by the Upper New Red, with beds of pebbles, some of which are derived from the Devonian and even from the Silurian. In this formation, near Sidmouth, have been found the remains of two remarkable reptiles, the *Hyperodapedon*, a strange form allied to the existing tuatera lizard of New Zealand and in England only known elsewhere in the formations of Warwickshire, and the *Labyrinthodon*, so named from the

Footprints of *Cheirotherium*, New Red Sandstone

intricate structure of its teeth, and also called *Cheirotherium*, from the hand-like impressions of its feet.

The Rhaetic beds are not well seen in Devonshire. They occur on the coast between Lyme Regis and the mouth of the Axe, and in the estuary of that river, but are much hidden by landslips of cretaceous formations from above. One layer, consisting of black shale, with bivalve shells such as *Cardium* and *Pecten*, contains also a bone-bed, with remains of fish, such as *Acrodus* and

*Hybodus.* The former is represented by its blunt teeth, and the latter, which was a huge, shark-like creature, by its long and formidable-looking fin-spines.

The Lower Lias is exposed in a narrow strip of coast from the Devonshire border to the mouth of the Axe, and to a greater extent in the valley of the river above Axminster. It has been divided on the coast into four distinct zones, each characterised by its own particular species of ammonite.

The cretaceous formations occupy a much wider area, but they also are confined to the southern part of the county. The Greensands of the Blackdown and Haldon Hills have been divided by geologists into fifteen layers, varying in thickness from a few inches to as much as thirty-five feet, some with few fossils, and some very rich in animal remains. *Trigonia* and *Inoceramus* are found in almost all the zones : other forms less widely distributed are *Murex* and *Turritella.* Chalk occurs on the south coast from the Dorset border to Sidmouth; and in isolated patches it extends inland as far as the Blackdown Hills, and also further west, in the Haldons. The Lower Chalk, well seen on the coast and to the west of Hinton, is made up of calcareous sandstones, with ammonites and pectens. The Middle beds, composed of white chalk with flints, the zone of *Terebratulina gracilis*, is exposed at Beer. The lower and harder. layer is characterised by *Rhynconella.* The Upper Chalk also holds many flints, with echini ; *Holaster* in the lower, and *Micraster* in the upper strata.

Last of all come the tertiary deposits, which, however,

occupy only a small area in the south-east, chiefly in the valley of the Teign, from Kingsteignton to Bovey Tracy; and there are a few isolated patches, as for example near Bideford and at Plymouth. These beds consist of clays, some of them of much value, with flints from the chalk, and gravels and beds of sand derived from the wearing away of older rocks. The most interesting feature of this formation is the lignite of Bovey Tracy, on the eastern edge of Dartmoor. Lignite, otherwise known as brown coal, consists of the imperfectly fossilised remains of tropical or sub-tropical vegetation, such as the palm, cinnamon, and laurel, amongst which are found lumps of resin. By far the most abundant remains are those of a very large tree allied to the sequoia of California. It is very remarkable that in the Pleistocene clay above the lignite are found stems and twigs of Arctic birch and willow, suggestive of a far colder climate than prevailed in Tertiary times, when the trees that went to form the lignite were growing.

To the Pleistocene period also belong the gravels and alluvial deposits of some of the river valleys (those of the Exe and the Teign, for example), the blown sands of Braunton Burrows and elsewhere, the raised sea-beaches, the submerged forests, and the cave-deposits which are alluded to in other chapters.

## 7. Natural History.

It is generally believed by naturalists that the ancestors of most of our fauna and flora reached this country at a time when what we now call the British Isles formed part of the mainland of Europe, and when there was no intervening sea to bar the way.

Before this colonisation was complete, however—that is, before all the different kinds of European beasts and birds had made their way to the extreme western districts—communication with the continent was broken off. The land of the north-western districts of Europe sank. The sea flowed in, forming the German Ocean, the English Channel and the Irish Sea, and the influx of animal life was stopped.

This is the reason why there are more than twice as many kinds of land animals in Germany as there are in England, and nearly twice as many in England as there are in Ireland. This is the reason why there are no snakes in Ireland, and why the nightingale, on returning from the south, never crosses into the sister kingdom.

On islands that have long been separated from a continent it is found that forms of life tend to vary in the lapse of time, and that fresh species are developed. That it is not long, as geological periods go, since Great Britain became an island, is shown by the fact that we have no quadruped or reptile except the Irish weasel (*Mustela hibernica*), and, setting aside minor differences which some writers have magnified to the value of a species, only one bird, the red grouse, which is not also to be found in Europe. Very different is the case in Japan, which was separated from the mainland of Asia so long ago that new species have had time to develope ; and the islands of that country contain many kinds of beasts and birds which are unknown on the adjacent continent.

Some of the animals which came from Europe into Britain have died out, either because the climate changed and so cut off their food supply, or because they were destroyed by the hunters of the Stone Age. The bones which have been found in Kent's Cavern at Torquay, and in other caverns, afford clear evidence that the mammoth, the lion, the bear, and the hyaena once roamed over the hills of Devonshire.

Although there are many more species of beasts and birds on the continent of Europe than there are in this country, both birds and beasts are numerically much more common here. Nothing strikes a naturalist more forcibly when travelling in France or Italy, for example, than the scarcity of wild life, and especially the fewness of the birds. It is true that we have fewer species, but we have many more individuals. To this, several causes have contributed. Englishmen do not, as is the custom in many European countries, shoot or trap for food small birds of every description. And game preserving—although it has been fatal to the larger birds of prey, such as kites, falcons, and buzzards, and keeps down other species, such as jays, magpies, and carrion crows—provides innumerable sanctuaries for great numbers of the smaller birds, which are safe from harm during the breeding season.

The natural features of Devonshire are so varied in character, including as they do large areas of wild and uncultivated and thinly-inhabited country, together with many well-wooded and sequestered valleys, and wide

A Red Deer

stretches of bog, salt-marsh, and sea-coast, that it is very rich in both animal and vegetable life. Its marine fauna and flora, in particular, are of very great interest, and are among the most remarkable in England.

Nearly all the native mammals of the British Isles are found or have been found in this county, from the "tall red deer" that has run wild on Exmoor from time immemorial, down to the pygmy shrew, the smallest but one of European quadrupeds, and weighing only one-tenth of an ounce, or about forty-three grains and a half.

Otters

Among the eight species of Devonshire bats is the very rare particoloured bat (*Vesperugo discolor*), of which the only example ever recorded in England was taken at Plymouth, having perhaps travelled there in the rigging of a ship. It is probably more than a hundred years since the last genuine wild-cat was seen in the county, but both the marten and the polecat still survive in secluded spots. Foxes are common, and there are still many badgers in some of the

Dartmoor valleys, where the two species have been known to inhabit the same holt. Otters abound on all the principal streams, and are as regularly hunted as the red deer and the fox. Devonshire is, indeed, pre-eminent for its otter-hunting, and the Culmstock pack is believed to be the oldest in the island. Harvest mice and dormice, although widely distributed, are not numerous, and the original English black rat is now rare.

Among the many marine mammalia that have been recorded for the county are two kinds of seal, the sperm-whale, the common rorqual—of which specimens nearly 70 feet long have been brought into Plymouth—the rare bottle-nosed dolphin and the still rarer Risso's grampus. Bones of a whale called *Balaenoptera robustus*, which were once washed ashore in Torbay, are said to represent a species so rare that these and a few similar relics stranded in Sweden are the only remains of it that have ever been found.

Situated as Devonshire is, between the English and the Bristol Channels, and containing widely-different physical features, suited to the needs of species of very different habits, the list of its birds, including residents, migrants, occasional visitors, and stragglers from the Atlantic and even from America, is a very long one.

Among the larger land-birds which still hold their ground in the county are the raven and buzzard, both of which are to be seen on Exmoor and Dartmoor and on the coast, and the peregrine falcon, which has eyries on both the northern and southern seaboards. A few pairs of choughs still build in the northern cliffs; while such rare birds as Montagu's harrier—first identified as a British species in this county—the hoopoe, and the golden oriole still occasionally breed here, and might do so regularly were they left in peace. Several birds, such as the kite and the osprey, the latter of which now breeds nowhere in England, and the former only in one solitary spot, have long since left the county. Warblers as a family are less abundant than in some other parts of the British Isles. The nightingale is nowhere common, but it occurs every season near Ashburton and in the valley of the Teign. Owing to the mildness of the climate it is not at all an unusual thing for a few chiffchaffs and willow-warblers to spend the winter in sheltered valleys on the south coast, instead of migrating to Africa in the autumn. The ring-ouzel is a regular visitor to the open country of Dartmoor, while the dipper haunts many of its streams. Two birds which have greatly increased in numbers of late years are the jackdaw and the starling. It is thought that the former has done much towards exterminating the chough by destroying its eggs; and the latter, by taking possession of its holes, has in many places driven away the green woodpecker. Partridges and pheasants are numerous, but black-game, once abundant on Dartmoor, have become so scarce that they are at present protected the whole year round.

But by far the most abundant, and perhaps the most characteristic, of the birds of Devonshire are the sea-fowl, the water-fowl, and the waders, of which more than 140 different kinds have been recorded for the county. Not only are its sandy shores, its bays and estuaries and leys, haunted in autumn and winter by multitudes of northern immigrants—swans, geese, ducks and a great variety of wading-birds ; but there are several spots along the south coast and a few on the north where sea-birds regularly breed ; while the reed-beds of Slapton Ley provide sanctuary for great numbers of coots and for many wild-ducks and teal, together with some rarer species. Herons are common on the south coast and along the river estuaries, and there are heronries at Powderham and elsewhere. A great black-headed gull (*Larus ichthyaetus*), shot on the Exe in 1859, is the only one known to have been seen in the British Islands.

There is, however, nothing on the mainland of Devonshire to compare in ornithological interest with Lundy, which in the summer time is a bird-lover's paradise. Gannets, once very numerous, have now left the island, but cormorants, shags and gulls of various species here build their untidy nests. Here multitudes of guillemots and razorbills assemble in the spring and lay their great pear-shaped and boldly-marked eggs on the ledges of the cliffs ; while even vaster hosts of puffins come back every year to take up their quarters in rabbit-burrows or in holes which they have dug for themselves in the turf. Here the raven, the buzzard, and the peregrine have fastnesses. Here, in chinks and crannies, storm-petrels breed ; and here, when darkness falls, the startled listener may hear the weird, wailing cry of the night-wandering shearwaters.

The few reptiles and batrachians of Devonshire present no points of special interest. Vipers abound on Dartmoor, where they are commoner than grass-snakes.

It is curious that, while the palmated newt is common throughout the county, the smooth newt and the triton are now comparatively rare.

The freshwater fish differ little from those found in the neighbouring counties; but there are fewer kinds in Devonshire than there are in the midlands or in the east of England. Trout abound in all the streams, and there are important salmon-fisheries on the Exe, the Dart, and other rivers. A sturgeon seven-and-a-half feet long was once taken in the Exe. Eels, which are hatched in the Atlantic, to the west and north of the British Islands, at a depth of 3000 feet or more, come up from the sea when they are two years old, and still very small, and ascend the rivers, especially the Exe, in enormous numbers. When they are mature, which is not until they are several years old, they go down to the sea to spawn, and never return.

It is, however, in marine zoology, for which few other parts of England afford so rich a field, and for which its bays and inlets, its rock-pools and stretches of sand provide ideal hunting-ground for the naturalist, that Devonshire is most distinguished. Many famous zoologists, such as Leach, Montagu, Parfitt, Gosse, and Kingsley have won renown both for themselves and for the county by their researches; while the Marine Biological Laboratory at Plymouth is constantly adding to our knowledge of the multitudinous inhabitants of the sea. The subject is so vast that only a few chief points can here be touched upon.

The sea-fish differ in marked degree from those of the

east coast of England. Plaice and cod, for example, are smaller here than those caught in the North Sea and the latter are scarce; and the haddock, one of the most important of east coast fish, is here almost unknown. Two characteristic fish of the south coast of Devon are the pollack, which reaches a great size, and the pilchard, confined to this county and to Cornwall. Many southern and even Mediterranean species find their way to these waters: notable examples are the gigantic tunny, one specimen of which weighed 700 pounds, the beautiful rainbow wrasse, one of the most brilliantly-coloured of all fish, and the boar-fish, which is sometimes quite common. A number of rare species, such as Montagu's sucker and the crystal goby, were first made known as British through being taken off the Devonshire coast. Stray examples of the tropical bonito, the flying-fish, the electric torpedo, and the sun-fish, one specimen of which weighed 500 pounds, and the splendidly-coloured opal or king-fish, have been recorded. Several kinds of sharks have been caught in these waters, including the blue shark, the spinous shark, covered all over with sharp prickles, the rare and formidable hammer-head, the huge thresher, and the still larger basking-shark. The latter is, indeed, the largest of British fish. Specimens have been caught measuring 30 feet in length, and weighing more than eight tons. Marketable marine-fish will be treated of in a later chapter.

Rich as are the Devonshire seas in fish, they are richer still in crustaceans—crabs, lobsters, prawns, shrimps and their allies; and in this respect ours is the premier county of England. Among a multitude of species, two which have occurred nowhere else in Britain may be specially singled out. One of these is the burying-shrimp, *Callionassa subterranea*, a little creature something like a very small lobster, with one claw—sometimes the right and sometimes the left—very much larger than the other. It was one of Montagu's many discoveries, and was found two feet deep under the sand of the Kingsbridge estuary. The other rare species is the turtle-crab, *Planes minutus*, a few specimens of which have been drifted ashore on fronds of Sargasso weed. The "small grasshoppers" which Columbus saw floating in the sea a few days before he sighted the New World, were, it is believed, not grasshoppers, but turtle-crabs.

Other and very beautiful forms of marine life, such as starfish, anemones, corals and other zoophytes, and sea-shells are very abundant. And in spite of the comparative scarcity of lime in the soil of Devonshire, the list of land and freshwater shells is a long one. It is remarkable that *Limnaea stagnalis* and *Planorbis corneus*, two water-shells that are common in Somerset, are unknown in Devon. The pearl-bearing mussel, *Unio margaritifer*, is found in both the Taw and the Teign.

The county is rich in insects, especially as regards butterflies, moths, and beetles; but several of the first-named which have been caught in Somerset have not been recorded here. The black-veined white (*Pieris crataegi*), once a common insect, has disappeared within the last forty years, and the greasy fritillary (*Melitaea Artemis*)—another vanishing species—is now almost extinct. Neither

insect can have been hunted down for the sake of its
beauty or its rarity, and the reason for this disappearance
is unknown.

As in the case of birds, the county is, from its position,
a favourite alighting-place for insects coming from abroad.
Between 1876 and 1890 large numbers of a very striking
and beautiful American butterfly, *Danais plexippus*, ap-

**Spurge Hawk Moth, with Pupa and Caterpillar**

peared in England, having apparently crossed the Atlantic,
and three specimens were caught in Devonshire. The
Lulworth skipper (*Hesperia Actaeon*), a small butterfly
which elsewhere is only found in Dorset, occurs along
the south-east coast of this county. Moths are very
abundant, and the first recorded British examples of
several species were taken in Devonshire.

About a hundred years ago, caterpillars of the spurge

4—2

hawk-moth (*Deilephila euphorbiae*) were very plentiful on spurge plants growing among the sand-hills near Barnstaple. Many of these caterpillars were taken by naturalists, and were reared, and ultimately turned into perfect insects ; although neither there nor anywhere else in our island was a wild example of this very beautiful moth ever seen alive. The spurge plants were long ago covered up by drifting sand, and the caterpillars were all destroyed. No other locality for them has been found in England, and as far as this country is concerned the spurge hawk-moth appears to be extinct.

As might be expected in a district of such varied physical features, with so mild a climate and such an ample rainfall, the flowering plants of Devonshire are very numerous, no fewer than 1156 species having been recorded. The abundance and beauty of its wild-flowers is one of the characteristics of the county. No one who has ever seen them will forget the wonderful wealth of primroses in some of the river valleys—at Holne, for example—or the splendour of the ling-empurpled sweeps of Dartmoor, or its sheets of golden gorse ; or the marvellous mist of bluebells upon woodland slopes or in the shelter of straggling hedgerows. Each several district, sea-shore and salt-marsh, moor and bog, wood and valley, has its own distinct and characteristic flora. One Devonshire plant, the Romulea or gênotte, *Romulea columnae*, a Mediterranean species with very small pale blue flowers, is abundant on the Warren at the mouth of the Exe, but grows nowhere else in England, although it is found in Guernsey. Several plants occur in only one other English county ; such for instance are the white rock-rose, *Helianthemum polifolium*, and the Irish spurge, *Euphorbia hibernica*, which are confined to Devon and Somerset, and the " flower of the Exe," *Lobelia urens*, which grows only in Devon and Cornwall. Three plants, which are very abundant in Somerset, the cowslip, the sweet violet, and the mistletoe, are rare in this county, although not unknown. The first plants of sea-kale ever brought into cultivation were originally dug up on Slapton sands ; and the vegetable came into note in Bath about 1775.

Ferns are characteristic of Devonshire. Not only are most of the familiar kinds abundant, but rarer species as the true maiden-hair, two filmy ferns, and the parsley fern (*Cryptogramme crispa*) are to be found. The magnificent royal fern, *Osmunda regalis*, still grows in some of the river valleys, and especially in Holne Chase, but it has suffered much from the greed of collectors, and the raids of unscrupulous dealers. A great variety of spleenworts has been recorded for the county, and one of the characteristic hedgerow ferns is the pretty little *Asplenium adiantum-nigrum*. Mosses, also, are very abundant, and there is one kind which occurs nowhere else in Britain. In sea-weeds Devonshire is richer than any other county except Dorset. Among its 468 different species is the Sargasso or Gulf-weed, sprays of which are sometimes thrown ashore after rough weather.

Except on the moors Devonshire is well timbered. The elm is perhaps the most conspicuous tree, but the beech and the ash are also very abundant. There is a very fine wych-elm, with a trunk 16 ft. in circumference, in Sharpham Park. The sycamore, which when well-developed is a very beautiful tree, here attains to fine proportions, and there are noble examples at Widecombe-in-the-Moor. The oak, although it grows freely, does not, as a rule, reach a great size, though there are some well-grown specimens at Tawstock Court. There is an oak at Flitton, near North Molton, which is thirty-three feet in circumference, and the Meavy oak is twenty-five feet in girth. An oak-tree thirteen and a half feet in diameter was cut down at Okehampton in 1776, and there is a tradition that two couples danced upon its stump. There are no very remarkable yews in Devonshire. Probably the finest are at Stoke Gabriel, Kenn, and Withycombe Raleigh, but the first of these is only fifteen feet in girth at the level of the ground. There is a story that, under the yew-tree at Mamhead, Boswell vowed that he would never get drunk again. At Bowringsleigh there is a magnificent avenue of lime-trees, and the avenue of araucarias at Bicton, planted in 1842, is said to be the finest in the kingdom. Several manor-houses possess one or more noble old mulberry-trees planted in the time of James I, with a view to encourage the cultivation of silk. At Buckland Abbey, once the home of Sir Francis Drake, there are some beautiful tulip-trees. Palms and other sub-tropical trees grow without protection at several places on the south coast; and at Kingsbridge and other towns pomegranates, oranges, lemons, and citrons will ripen their fruit in the open air.

A good many places in Devonshire take their names

from trees. Thus Ashburton is named from the ash,
Egg Buckland from the oak, Bickleigh from the beech,
and Holne from the holly.

## 8.  A Peregrination of the Coast: 1, The Bristol Channel.

Devonshire, like Cornwall and Kent, is remarkable
in having both a northern and a southern seaboard ; a
peculiarity shared by no other English county.  Its two
shores present striking points of difference.  The south
coast-line is broken by many estuaries.  On the other
shore there is only one important river mouth.  There
are, it is true, many little coves and inlets on the Bristol
Channel, some of them of great beauty ; but they make
little show upon the map of England, and the stern
outline of the North Devon coast affords no harbour of
refuge.

Both shores are rock-bound.  But while the southern
cliffs are, in great measure, of warm-hued and even
brightly-coloured stone, those on the north are dark
and gloomy; and their tones, although in some places
very beautiful, are set in quieter key—in grey or brown
or even verging upon black.  Again, the southern shore
is fringed at some points with sandy beaches ; while on
the north coast there are no sands at all, except on the
western side of Bideford Bay.

Along the northern seaboard of Devon there runs
a series of magnificent cliffs, in parts heavily wooded,

whose dark walls, sloping steeply to the shore and with projecting bases suggestive of the ram of a battleship, are relieved at many points by deep, rocky clefts, known variously as combes or mouths; each with its stream, each green with ferns and oak-coppice and thickets of thorn and hazel, and each with its butterfly-haunted clumps of tall hemp-agrimony.

The Castle Rock, Lynton

Down such a hollow, the deep and finely-wooded valley of Glenthorne, runs the border-line that divides Somerset from Devon. Rather more than three miles west of it there stands out into the Bristol Channel the dark mass of Countisbury Foreland, the most northerly point in the county, and one of the highest along its coast, 1100 feet above sea-level. Four miles beyond the Foreland, at the mouth of a deep and well-wooded valley, down which runs the beautiful trout-stream from which it takes its name, is Lynmouth, famous for its scenery, of which two striking features are the Watersmeet on the river, and the Valley of Rocks on the coast. A port and fishing-village up to the close of the eighteenth century, its small tidal harbour is visited now only by a few small coasting vessels.   About four miles west of

Valley of Rocks, Lynton

Lynmouth is Heddon's Mouth, a little bay at the foot of towering cliffs, with another trout-stream flowing down to the sea through one of the loveliest combes in North Devon.   Five miles of cliff stretch from Heddon's Mouth to Combe Martin Bay, a little inlet lying in the shelter of two conspicuous heights, the Great Hangman and the Little Hangman—names associated with no tragic story, but derived, like many others round our coasts, from the Celtic *maen*, a stone—and with its village, once famous for its rich silver-mines, running a mile inland. Two miles of rock-bound and dangerous coast, swept, especially off Rillage Point, by a strong tide-race, extend from Combe Martin Bay to the ancient port of Ilfracombe, whose mild yet bracing climate and beautiful surroundings have made it the most popular seaside resort in North Devon. Its little land-locked harbour is almost surrounded by lofty hills and rugged cliffs, whose beauty is greatly heightened by the varied colouring of the rock and by the vivid green of the abundant vegetation.

Ilfracombe is a place that has played a part in history. In the fourteenth century it provided six ships towards Edward III's expedition against Calais. It was from this port that Queen Elizabeth sent troops to Ireland during the rebellion of the Earl of Tyrone. In the Civil War it was taken alternately by Royalists and Parliamentarians. It was from Ilfracombe that Wade and Ferguson and other Sedgemoor fugitives tried in vain to escape by sea. And it was here, in 1796, that the French squadron which afterwards landed 1000 scoundrels of the *Légion noire* at Fishguard, on the opposite coast—the last hostile invasion of these islands—burnt the fishing-smacks lying in the harbour. The French ships were in the end taken by Lord Bridport.

A short distance west of Ilfracombe is Wildersmouth, a beautiful bay, with a gravelly beach, famous for its richness in the lower forms of marine life, and three miles farther down the coast juts out Bull Point, a bold headland guarded by a powerful lighthouse, marking the north-

Ilfracombe, from Hillsborough

eastern limit of the most dangerous part of the coast, which here turns abruptly southward, facing squarely to the open Atlantic. A little farther on is Morte Point, whose name the popular fancy regards, although without foundation, as hinting at the deadly character of its black, jagged, sea-swept rocks. The village of Mortehoe, a few hundred yards inland, was the property in the thirteenth century of the de Traci family, one of whom was among the murderers of Thomas à Becket. But there is no ground for the legend that he was buried here, or for the traditions of him that are current in the district. A tiny little cove on the south side of Morte Point, called Barracane Beach, was once famous for its rare and beautiful shells ; but it is now so widely known, and its charm is so completely lost, that it has been said of it that there are more collectors than specimens.

Beyond Morte Point is Morte Bay, most of whose shore lies low, and is fringed throughout almost its entire length by the broad expanse of Woollacombe Sands, along whose margin, at heights varying from eight to fifteen feet above high-water mark, may be traced at intervals a raised sea-beach. At the southern extremity of Morte Bay is the noble headland of Baggy Point, a magnificent piece of cliff, haunted by crowds of sea-birds, and pierced by many caves. The shore of Croyde Bay, beyond the Point, is famous for its fertility ; and from the crest of Saunton Down, the last headland before the estuary formed by the waters of the Taw and the Torridge, is a view which, embracing sea and coast-line, rich expanses of farm-land, the distant heights of Dartmoor

and the faint shape of Lundy on the far horizon, is one of
the finest in all Devon.  Along the shore to the south
of Baggy Point, where Saunton Sands form the seaward
fringe of Braunton Burrows, is another long stretch of
raised sea-beach, from two to fifteen feet above high-
water mark.  And in this beach, not far from Saunton,
is a large boulder of red granite, a rock unknown in
the district, which may have been stranded here by
floating ice.

Braunton Burrows is a long, wide tract of sand-hills,
some eighteen square miles in area, stretching far inland,
and reaching to the estuary of the Taw and the Torridge,
with deep hollows among which, without a compass, it
is quite possible to get completely lost.  It is a place of
much interest to the naturalist and the antiquarian.  A
number of rare plants are found here, great quantities
of primitive flint implements have been discovered in the
sand, and at low water the remains of a submerged forest
are to be seen along the shore.

The estuary formed by the combined streams of the
Taw and Torridge, the former of which is also known
as the Barnstaple River, flows into Barnstaple Bay at the
south end of Braunton Burrows.  There is no port on
the open coast ; but just inside the estuary are the quaint
old town of Appledore and the equally ancient village of
Instow, on the left and right banks, respectively, of the
river Torridge.  In the mouth of the same stream, a
little to the south of Appledore, is a long flat rock called
the Hubblestone; named, according to tradition, after the
viking Hubba, who pillaged this coast in the reign of

King Alfred, and fell in battle at the mouth of the Parrett, in the adjoining county of Somerset.

Blocking up a great part of the river mouth, and stretching down the coast past Westward Ho! a distance of about two miles, is the Pebble Ridge, a remarkable bank of shingle and sea-worn boulders, some of which are of great size, though the majority are not more than a few inches in diameter. The sea has gradually shifted it further and further inland, and it now covers what was once a long stretch of good pasture-ground. On its landward side are the golf-links of Northam Burrows, considered to be among the finest south of the Tweed.

Westward Ho! a modern watering-place named in honour of Kingsley's great romance, is chiefly interesting on account of its submerged forest, in whose peat and clay, deeply covered by the sea at high tide, have been found, not only the trunks of large oak and fir-trees, and bones of the wild boar, stag, horse, and dog, but bones of man, together with charcoal, pottery, and implements of flint.

Six miles south-west of Westward Ho! and in the centre of the curve that marks the southern shore of Barnstaple Bay, is the prettily situated fishing-village of Buck's Mill, with red and wood-crowned cliffs behind and beyond it, and extending to Clovelly, the famous little town that may truly be called one of the most remarkable spots, not in Devonshire only, but in all England. Crowded in a hollow in the cliff, with woods on either side, and with an air of climbing up from its little tidal harbour sheltered by a rough stone pier of the time of Richard II, it consists of one long, winding, pebble-paved street, too steep for wheeled traffic, with quaint and irregularly-built cottages to left and right, beautiful with creepers and myrtles, fuchsias and geraniums. Not only is Clovelly intimately associated with the memory of Charles Kingsley, whose father was rector here, but it

Cliffs near Clovelly

is the original " village of Steepways," in Dickens and Collins' Christmas story, *A Message from the Sea.*

A long stretch of wild and magnificent coast-line extends from Clovelly to Hartland Point, where the shore again turns southward, and again from Hartland to the county border ; a wall of precipitous black cliffs, relieved here and there by bands of red schist, and broken at intervals by green combes such as are characteristic of the seaboard of Devon ; a terrible coast, strewn with fragments of wreckage from ill-fated ships.

Hartland Point, believed to be the Promontory of Hercules alluded to by the geographer Ptolemy, is a noble headland, whose dark steeps rise 350 feet sheer up

Clovelly Harbour

out of a dangerous and ever restless sea. Perhaps there is not, in any other part of North Devon, more striking evidence of volcanic upheaval and disturbance than is to be seen in the curved and gnarled and twisted strata of the cliffs that tower above Hartland Quay.

Six miles south of Hartland the northern seaboard of the county ends, as it began, in a deep hollow in the cliffs, Marsland Mouth, a beautiful combe, down which, under storm-beaten oaks and thickets of thorn and hazel, there winds the stream that forms the border-line between Devonshire and Cornwall.

Church Rock, Clovelly

## 9.   A Peregrination of the Coast : 2, The English Channel.

The points that specially characterise the southern seaboard of Devonshire, and distinguish it from the northern shore, are its many estuaries, its numerous bays and bold headlands, the strong, deep red, in some places, of its rugged cliffs, and, in a minor degree, the sandy beaches which lend an added charm to many of its seaside towns.

No natural feature marks the spot, half-way between Lyme Cobb and the Seven Rocks Point, where the border-line between Dorsetshire and Devonshire begins. But all that part of the coast, almost as far as the mouth of the Axe, shows signs of having been broken away by repeated landslips; one of the most serious of which happened in 1839, when a vast mass of cliff, extending all the way from Pinhay (or Pinner) to Culverhole Point, slipped bodily down some 300 feet, carrying with it fields and houses; and it now lies in most picturesque ruin on the beach.

The mouth of the Axe, above whose eastern side rises the Haven Cliff, a fine mass of red sandstone crowned by white chalk, has long since been silted up by pebbles, and no ships now visit either Axmouth or Seaton, the latter of which was once of sufficient importance to contribute two vessels towards Edward III's expedition against Calais, but is now only a watering-place. Beyond the mouth of the Axe, separated from it by a mile of low-lying shore, the White Cliff, also a scene of many landslips, rises sheer up out of the sea; a fine piece of cliff-wall, the effect of whose bands of red and white, of brown and grey, is greatly heightened by the green of its abundant vegetation. More striking still is the white precipice of Beer Head, the most southerly outcrop of chalk in England, worn above into picturesque and ivy-mantled crags, and hollowed at its base into many caves. From its summit, 426 feet above the sea, is a far-reaching view of the coast, covering the 50 miles from Portland on the east to the Start on the west. Half-

Pinhay Landslip

way between the mouth of the Axe and Beer Head is
the quaint and old-world village of Beer, famous for its
labyrinthine quarries tunnelled deep into the hill, for its
fisheries and lace-making, and, formerly, as a special haunt
of smugglers.   From Beer Head, past the little openings
of Branscombe Mouth, Weston Mouth, and Salcombe,
to Sidmouth, is a range of magnificent and picturesquely-

White Cliff, Seaton

coloured cliffs, white and grey and yellow, and at some
points rising straight up from the sea-line.

Sidmouth, the "Baymouth" of Thackeray's *Pendennis*,
set among beautiful hills, and one of the pleasantest of
west-country watering-places, was once a port, with
valuable pilchard fisheries.   But its harbour has been
destroyed by repeated falls of rock from its grand cliffs
of deep red standstone, the Sid is silted up with sand and

shingle, and the pilchards have left this part of the coast.
About a mile west of Sidmouth is the beautiful headland
of High Peak, whose summit, 511 feet above the sea, is
the most lofty point on the south coast of Devon.   Just
beyond it is the popular bathing-place of Ladram Cove,
whose firm sands are fringed with brightly-coloured
pebbles.   Rather more than two miles farther on is the
estuary of the Otter, a harbour 500 years ago, but now,
like so many of these river mouths, barred with shingle.
Close to the estuary lies the quiet little town of Budleigh
Salterton, set in a beautiful valley, famous for its mild
climate and its luxuriant vegetation.   Some five miles of
coast-line—broken half-way by Straight Point, beyond
which the shore is low—extend from Budleigh to the
mouth of the Exe, the widest of Devonshire estuaries,
but almost closed by a long bar of grass-grown sand
called the Warren, on which, during the Civil War,
stood a Royalist fort mounting sixteen guns.   Exmouth,
at the east side of the estuary, formerly a fishing-village,
is now a highly popular watering-place.

Four miles farther on, in a little bay walled-in by
lofty cliffs of deep red sandstone, is Dawlish, noted for its
warm climate and its good sands.   At the eastern end
of the bay is a rock called the Langstone, and at the
western end are the strange-looking pillars of red sandstone known as the Parson and Clerk.   Teignmouth lies
rather more than two miles S.S.W. of Dawlish, with
picturesque red cliffs and firm sands all the way, at the
mouth of the estuary of the river Teign, whose swiftly-
flowing stream is here crossed by one of the longest

Parson and Clerk Rocks, Dawlish

wooden bridges in England. It is a small port and a very popular watering-place, with beautiful inland scenery behind it, and inside the Den—the dune or sand-bank which bars a great part of the river's mouth—is a good harbour for vessels of light draught. Teignmouth is one of the towns that in the past have suffered from the attacks of the French, who burnt it in 1347 and again in 1690.

Anstis Cove, near Torquay

Four miles south of the estuary of the Teign is Babbacombe Bay, in whose beautiful cliffs of red and grey is some of the richest colouring on the whole coast. The paler-toned cliffs round the picturesque little inlet of Anstis Cove are of limestone. Half a mile farther is the prominent cape called Hope's Nose, the northern limit of Torbay, and a spot of much interest to the geologist on account of the raised sea-beach which, at a height of some thirty feet above the present high tide-line, may be traced under the headland, and also, at a lower level, on the Thatcher Rock. Among the marine shells of the latter deposit is *Trophon truncatus*, an arctic species, whose presence here is another proof that the climate of Devonshire was once far colder than it is now.

Torquay from Vane Hill

Torbay, which extends from Hope's Nose on the north to Berry Head on the south—two prominent head-lands nearly five miles apart—is one of the best known and most beautiful bays on the coast of England. In all except easterly winds it affords an excellent anchorage which was much used by ships of the Royal Navy in the old sailing days, and it is still a great yachting station. At the northern end of the bay, occupying, it is said, more ground in proportion to its population than any other town in the island, is the much frequented watering-place of Torquay, widely celebrated for the beauty of its situation and the mildness of its winter climate. Along the whole coast of Torbay, at a level which shows that the land has sunk some forty feet, lies a submerged forest, in which have been found bones of the wild boar, red-deer antlers, and mammoth's teeth. But proofs of an elevation on a still greater scale are to be found in the borings of sea-shells in the limestone cliffs above Kent's Cavern, within the limits of the town, at a height of 200 feet above the present sea level.

Half-way along the shore of Torbay is Paignton, another favourite seaside resort, famous for its fine beach, and on a steep slope at the head of an inlet rather more than a mile before coming to Berry Head stands Brixham, a town second only in importance to Plymouth among the fishing-stations of the south coast of England. Here, on the 5th of November, 1688, the Prince of Orange landed. And here, six weeks after the Battle of Waterloo, the *Bellerophon* anchored, with Napoleon Buonaparte a prisoner on board.

Beyond Berry Head, which forms the end of a broad promontory, worn at its base into many caves, and noted for its quarries, there extends for many miles— all the way, in fact, to the mouth of the Dart—a stretch of very beautiful coast-line, with low but finely-coloured cliffs of sandstone and limestone and slate, varying in tint from red to purple, and from brown to grey, with a series of sandy bays and fringed by outlying rocks, two of which are called Mewstones. One of these, standing just where the coast sweeps round to the estuary of the Dart, is a lofty pinnacle of stone more than 100 feet high. Well inside the mouth of the Dart, on the steep slope of its left or eastern entrance, is the quaint little town

Brixham

of Kingswear; and opposite to it, on the western shore, lies Dartmouth, once a noted port, but now only a favourite yachting station. The old man-of-war, the *Britannia*, anchored here close to land and long used as a training-ship, has been superseded by a naval college on shore, and is now used only as a store. Dartmouth is a place of much historic interest. It was from here that part of Richard Cœur de Lion's crusading fleet sailed

for Palestine. The port furnished thirty-one ships towards
Edward III's attack on Calais. Twice, in the half
century that followed, it was plundered by the French.
It played a prominent part in the Civil War, and was
taken first by Prince Maurice, and afterwards by Fairfax.

Between the mouth of the Dart and Start Point, nine
miles as the crow flies, is Start Bay, walled for about half

The " Britannia " and " Hindostan " in Dartmouth Harbour

its length with low and quiet-coloured cliffs of slate, and
fringed in great part with sand and shingle. At Blackpool, a picturesque little cove near the northern end of
the bay, du Chastel the Breton landed, in 1404, on a
pillaging expedition, for the plundering was not all on
the side of the English. But the Frenchman was killed,
with 400 of his men, and 200 more were taken prisoners.

Halfway along the shore of Start Bay are Slapton Sands, where a beach of small and brightly-coloured pebbles and a bank of shingle separate the long and narrow lake called Slapton Ley from the waters of the Channel. Off this spot, marked by two beacons on the shore, is the "measured mile" for testing the speed of steamships. Not far from Slapton the coast rises again, and above the fishing villages of Hallsands and Beesands, which stand at the water's edge, reaches a height of some hundreds of feet. The people of these two little hamlets train powerful dogs, which, in rough weather, swim out through the surf, catch the painters thrown to them and thus enable the fishing-boats to be dragged ashore.

Start Point, or, as it is perhaps more often called, the Start, is one of the famous capes of Britain, a bold headland sloping steeply both ways, like the roof of a house; whose iron base, fringed with white quartz pebbles, has been the scene of many shipwrecks, and whose dark cliffs and rugged crags are haunted by multitudes of sea-birds. The cliffs of this part of Devon, from the Start round Prawle Point and Bolt Head to Bolt Tail—cliffs whose grey rock, relieved by bands of white quartz, has been bent and twisted by volcanic upheaval, and weathered by rain and frost, by wind and sea, into the wildest and most fantastic shapes—are as remarkable for picturesqueness of form as other parts are for richness of colouring. Three miles beyond the Start is Prawle Point, a magnificent mass of jagged rock, the most southerly point in the county, and a well-known steering-mark for ships in the Channel. It was off this

shore, in 1793, that the English ship *Nymphe* captured
the French man-of-war *Cléopatre*; the first naval battle in
the struggle between England and the French Republic.
Between Prawle Point and Bolt Head is Salcombe Mouth,
a creek rather than an estuary; a long, winding, and
picturesque inlet, whose entrance is obstructed by a
bank of sand. Trunks of oak and other trees, from a
submerged forest not far from land, are sometimes thrown
ashore here after rough weather. To the west of Salcombe
stands Bolt Head, of no great height, but a noble mass of
rugged and weather-worn rock. Beyond the Head the
coast rises into steep and lofty cliffs, culminating in Bolt
Tail, close under whose eastern face, in 1760, the 74-gun
ship *Ramillies* was lost, with more than 700 of her crew.
A gun recovered from the wreck lies by the Hope signal-
station, on the height above. These cliffs have been
much broken away by landslips; and a series of fissures
called the Pits suggest that much more ground is still
to fall.

Round Bigbury Bay, of which Bolt Tail is the
eastern limit, is some of the most beautiful scenery of
this beautiful coast. A striking feature of the bay is a
great rock called the Thurlestone, an outlying mass of
red sandstone, conspicuous against the general greyness
of the cliffs, and pierced by a lofty archway, worn by
wind and sea. Two estuaries, the Avon Mouth and the
Erme Mouth, break the coast-line of the bay; and there
is a third, called Yealm Mouth, near the entrance of
Plymouth Sound, a couple of miles beyond the grand
slate headland of Stoke Point. Outside the Avon Mouth

is Borough Island, carpeted in spring-time with the beautiful blue of the delicate little vernal squill. The Erme, whose mouth is guarded by rugged cliffs of slate, is strewn with rocks and sandbanks; but the estuary of the Yealm is a fine sheet of deep, navigable water. Standing far out into Wembury Bay, at the mouth of the Yealm, is the third of the Mewstones, a rocky and beautiful little islet, nearly 200 feet high, and frequented, as its name implies, by many sea-gulls.

The Mewstone may be said to mark the eastern side of the entrance of Plymouth Sound, one of the best known, most important, and most beautiful bays in the kingdom. It is by nature fully exposed to southerly winds, and it has, in the past, been the scene of many shipwrecks. But the breakwater, which was built in the early half of the nineteenth century right across it, two miles south of Plymouth Hoe, with the special object of sheltering ships of the Royal Navy, now affords a safe and excellent anchorage. Nearer the shore is Drake's Island, now strongly fortified, but in Stuart times a State prison, where Lambert, one of the most distinguished of Parliamentary generals, spent the last eighteen years of his life.

At the head of the Sound, on its eastern side, is the inlet called the Catwater, the estuary of the river Plym, an important mercantile anchorage, protected by Batten breakwater. It was here that the English Fleet waited until the Spanish Armada, on its way up the Channel, had passed the entrance of the Sound.

Between the Catwater and the Hamoaze, the great

naval anchorage which extends from the Sound to Saltash
Bridge, are the "Three Towns," Plymouth, Stonehouse,
and Devonport, now joined into one by continuous
buildings, forming the busiest and most populous part
of the county, and constituting, with their dockyards,
barracks, gun-wharves, and victualling yards, one of the
most important stations of the Royal Navy.

## 10. Coastal Gains and Losses. Sand- banks. Lighthouses.

There are parts of our island where, even within
historic times, the coast-line has been greatly changed
by the encroachment of the sea, usually through the
wearing away of the cliffs along the shore. This is
especially the case on the eastern coast of England,
where, in the lapse of ages, villages, towns, and whole
manors have been completely swept away. The old
town of Ravenspur, for example, a place that in its time
rivalled Hull as a sea-port, is to-day a mere sandbank far
out from shore; and the sea runs twenty feet deep over
the once great shipping town of Dunwich, whose site is
now two miles from the land.

On the other hand, there are places where the reverse
has happened; where the shore has gained upon the sea.
The town of Yarmouth, for instance, stands on ground
that first became firm enough to build upon nine
hundred years ago. A large tract of land on the coast
of Carnarvonshire has, in times much more recent, been

reclaimed from the sea; and the day cannot be far distant when the mud-flats of the Wash will be under the plough.

Similar changes—changes resulting both from gain and loss—have happened and are still happening in Devonshire. Braunton Great Field, a rich tract of land some 300 acres in extent, cut up into hundreds of small freeholds, was, it is believed, reclaimed from the estuary of the Taw. On the other hand, the Pebble Ridge on the shore of Barnstaple Bay has been slowly driven inland by the force of the sea, and is said to have advanced 200 yards in the last fifty years, thus covering a long stretch of pasture-land under heaps of stones. Attempts have lately been made, by means of piles and groynes of timber, to stop its further movement.

Much more remarkable, however, and much more widely distributed, are the alterations that have taken place on the south coast of Devonshire, owing mainly to erosion of the cliffs and consequent landslips, and to the washing up, by strong currents, of vast quantities of sand and shingle. From the Dorset border westwards, especially between Pinhay Bay and Culverhole Point, in the White Cliff near Seaton, and at Beer Head, long stretches of cliff, undermined probably by streams and heavy rains, have fallen, sometimes in masses half a mile long. The old town of Sidmouth is now buried under the shingle, the cliffs that protected the harbour having been entirely washed away. At Dawlish, again, rather more than fifty years since, a mass estimated at 4000 tons fell bodily into the sea. Nor is the erosion the work of natural forces alone.  In 1897 immense quantities of fine shingle were taken from the beach at Hallsands, to make concrete for Keyham dockyard, with the result that the beach there has sunk twelve feet, that high-water mark is now much farther in-shore, and that many houses in the village have been swept away

A Rough Sea at Ilfracombe

by the sea, whose further inroads have at last been checked by means of massive walls of concrete.

Most of the south coast estuaries, as has already been pointed out, have been more or less blocked up by banks of sand or shingle, some of which are still undergoing change.  The Warren, for example, the great bar at the mouth of the Exe, now connected with the western shore, was in the seventeenth century joined to the

K. D.                                                    6

Exmouth side of the river, and was still reached from there by stepping-stones as late as 1730. The Warren is now being slowly washed away, at the rate, it is said, of an acre in a year ; and the river has within historic times encroached upon the site of Newenham Abbey.

Great as has been the loss of land on the south coast, there have been some gains. More than 170 acres of land, for instance, have been reclaimed from the Laira near Plymouth ; and the village of Penny-come-quick, lower down, whose anglicised Celtic name means "the house at the head of the creek," is no longer at the water's edge. In 1805 some thirty acres were recovered from the Charleston marshes, on the Salcombe estuary.

The Bristol Channel is one of the most stormy and dangerous parts of the British seas, and is the scene of about one-tenth of all the shipping disasters that happen on our coasts every year. In its upper reaches navigation is made difficult by banks of mud and sand which are continually altering in shape and position. On the north coast of Devonshire, however, there are no out-lying sandbanks. There is a small patch of sand off Lynmouth, 1½ miles N.N.W. of Countisbury Foreland, and the estuary of the Taw and Torridge is obstructed by a dangerous and shifting sandbank known as Barnstaple Bar, upon which ("the harbour bar" of Kingsley's song), many vessels have been wrecked. But the dangers of this stormy shore lie mainly in the iron-bound coast itself, and in the rocks that stretch seaward from the bases of the cliffs. From Bull Point to Baggy Point, especially off Morte Point, and again from Clovelly to the border of Cornwall, particularly off Hartland Point, the shore is fringed with reefs and sharp edges of rock.

Lundy, again, is a constant source of danger to sailors; partly because of the many rocks that stretch out from it, especially the Hen and Chickens at the north end and the Lee Rocks at the south; partly because of the strong currents that, off the south point of the island, run five knots an hour; and partly because of the fogs that so frequently envelope it. It was all three causes combined that, in 1906, occasioned the loss of the first-class battleship *Montagu*, which, carried out of her course by the current, and deceived by the fog, became a total wreck on the Shutter Rock, the southern extremity of the island. Off Lundy, too, are the only banks of importance. Over the Stanley Bank, which lies to the north-east, where the depth at one point is only four and a half fathoms, there run, in heavy weather, the dangerous "tide-rips" known as the White Horses.

The navigation of the south shore of Devonshire is much more important than that of the north; partly because of the number of ports in the county itself, and partly because the English Channel is a much more crowded waterway.

The principal danger to navigation on the south coast is the group of reefs called the Eddystone Rocks, fourteen miles south-south-west of the entrance of Plymouth Sound. They are all covered at high tide, but the top of one of them is nineteen feet above low-water mark. In Plymouth Sound itself, especially near the eastern shore, there are many rocks and shallow patches. The most conspicuous of the former is the Mewstone, 194 feet high. On the Shagstone, a little farther in, the P. and O. steamship *Nepaul* was lost. From this point eastward the coast is fringed with rocks

The Eddystone Lighthouse

and small islets, most of them close in-shore. But from the Start a chain of sandbanks called the Skerries extends out some miles from the land. From Bigbury Bay to the Start is one of the most dangerous parts of the coast, and has been the scene of many wrecks. Here, to name a few of many instances, were lost the *Ramillies*, of whose crew 708 were drowned; the *Chanteloupe*, when only one man was saved; the *Marana*, from which seven men escaped; and the *Dryad*, on which every man perished.

From the mouth of the Dart to Hope's Nose there are many outlying rocks; but from that headland to the Dorsetshire border the coast is comparatively "clean," that is to say, free from obstructions. All the rivers east of Plymouth Sound are more or less blocked by bars, with the exception of the Dart, whose entrance, however, is strewn with rocks.

To warn the sailor against these and other dangers, buoys, bells, beacons, fog-horns or sirens, guns or explosive signals, and lighthouses have been provided at many points along the Devonshire coasts. There are also numerous storm-signalling stations, and there are no fewer than thirteen lifeboats, of which eight are on the south coast. The men of the thirty-three Devonshire Coastguard stations have been the means of saving many lives.

On the two coasts, including Lundy, there are in all fifty-one lights of various sorts and sizes, from the eight first-class lighthouses with massive stone towers, of which the most famous although not the most powerful is the Eddystone, down to the small but useful lights of a hundred candle-power or less, most of which are connected with quays and harbours; while others, like that at Clovelly, lighted only in the fishing-season, are temporary, a number of them consisting merely of a lantern on the top of a post. A hut that carries the red and white lights at the mouth of the Barnstaple river is on wheels, and is moved as the bar shifts its position.

The first Eddystone lighthouse, a fantastic structure of wood, with six stages, begun in 1690 by Winstanley— who, while engaged in building it, was carried off by a French privateer, but promptly released by command of Louis XIV—was swept away, with its builder and three other men, in the historic storm of 1703. The second lighthouse, also of wood, built by Rudyerd in 1706, was destroyed by fire in 1755. The third, which was the first real lighthouse ever erected, was constructed by Smeaton of stones dovetailed together, with a shaft eighty-seven feet high, shaped like the trunk of an oak-tree for the sake of strength, and with the idea that it would offer greater resistance to the waves. It was finished in 1759, but its woodwork having been burnt in 1770 was then replaced by stone. The foundations of this tower having been undermined by the sea, a fourth lighthouse, whose top is 133 feet above high-water mark, was built by Douglas, between 1878 and 1882, on a rock forty yards south-south-east of the original site, which is nine miles and a quarter from the nearest land. Part of the old tower was taken down, and re-erected on Plymouth Hoe, in memory of Smeaton. Like most of the Devonshire lights the Eddystone lantern is of the group-flashing order, giving a light equal to that of nearly 300,000 candles, with two quick flashes every half-minute, and visible in clear weather for seventeen miles. A minor fixed light, in the same tower, shines on the Hand Deeps, a bank to the north-west; and the lighthouse is also provided with an explosive fog-signal, giving two reports every five minutes.

Other very powerful lights, visible for from seventeen to twenty-one miles, are—giving them in ascending order—those at Hartland, Bull Point, Countisbury

The Start Lighthouse

Foreland, North Lundy, the Start, and South Lundy, the last named being the most brilliant of all, of 374,225 candle-power. All these lighthouses have fog-sirens or explosive fog-signals.

Most of the English lighthouses are under the charge of the Trinity House, a corporation founded in 1512, and now having a yearly revenue of £300,000, derived from "light-dues" levied on shipping. The expense of keeping up the lighthouses ot the United Kingdom in 1909 amounted, however, to £464,540. The early beacon-lights were simply fires of coal, and one of these was in use at St Bees Head as recently as 1812. There are now round the coasts of Britain more than a thousand lighthouses and lights of various degrees of importance, from those at the Lizard and at St Catherine's Point, which are the most brilliant in the world, and may be reckoned in millions of candles, down to insignificant little structures, of which there are many, like the 100-candle-power "Jack-in-the-Box" on the river Tees.

## 11. Climate and Rainfall.

The climate of any country, or in other words, its average weather, by which, again, we mean its temperature, rainfall, and hours of sunshine, as well as the dryness or otherwise of its air, depends upon various circumstances and conditions, but especially upon geographical position, that is to say, upon the nearness of the country to the equator, upon its distance from the sea and its height above sea-level; partly also upon its soil and vegetation. Speaking generally, the nearer we approach the equator the hotter will be the climate, and the nearer to the sea-coast the milder and more equable it will be. The highest temperature in the shade ever yet recorded, however—127° Fahr.—was in the Algerian Sahara, at a spot not even within the tropics; and the greatest cold ever experienced, 90° below zero, Fahr., was at a place in Siberia only just within the arctic circle.

The climate of the British Isles is very greatly influenced by the great ocean current, often called the Gulf Stream, from its supposed source in the Gulf of Mexico, which, impelled by steady winds, carries a constant stream of warm water from the equatorial regions towards the north pole. This current washes the shores of these islands, and it is this circumstance which makes our winters so much milder than those of Labrador, which is no nearer to the pole than we are. Were it not for the Gulf Stream our weather would probably be more severe than that of Newfoundland.

The highest shade temperature ever experienced in Britain was 101° Fahr., at Alton in Hampshire, in July, 1881 ; and the lowest was 10° below zero, Fahr., at Buxton in Derbyshire, in February, 1895. A temperature of 23° below zero is said to have been measured in Berwickshire, in 1879, but the correctness of the record has been questioned. What is, however, of more importance, is the average temperature for the year, which, for the whole of England, is 48° Fahr.

The sunniest part of England lies, as might be expected, along the south coast, although the south-east and south-west coasts also get more sunshine than inland districts in the same latitude. The sun is above the horizon in this country for more than 4450 hours in the year; but, owing to the frequent presence of clouds, he is not visible for even half that time in any part of the

British Isles and the average for the country as a whole is only 1535 hours. The southern coasts sometimes enjoy 2000 hours of bright sunshine in a year, but their average is probably not more than 1700 or 1800 hours, or about five hours a day all the year round. The amount decreases as we go north; while the manufacturing districts of the midland counties, owing to the smoke

The Winter Garden at Torquay

which so often obscures the sky, get 1200 hours or less. At Manchester in 1907 only 894 hours of sunshine were recorded for the year's total. The sunniest months throughout the country generally are May and June, and the gloomiest month is December.

The climate of a country, however, depends not only upon the sunshine, but upon the rainfall. The amount of rain that falls in any place varies according to the height of that place above the sea, its distance from the coast, and the configuration of the ground, that is to say, upon its position with respect to valleys, up which moisture-laden air may be driven by the wind, to be compressed and cooled until its moisture falls in rain.

The rainfall varies very much in different parts of England, but the average amount for the whole country is about 33 inches in a year. When we speak of an inch of rain we mean that it would lie an inch deep on a perfectly level piece of ground. Thus, if all the rain that fell in a year stayed on the ground, and did not evaporate, or run away, or sink into the earth, the water would be 33 inches deep all over England at the end of the twelve months. An inch of rain all over an acre of ground weighs rather more than 100 tons. Not only does the rainfall vary in different parts of the country, but it varies in different years. The wettest year on record was 1903, when the rainfall averaged 50 inches for the whole of England; and the driest year was 1887, when it amounted to no more than 24 inches.

The heaviest rainfall in Great Britain is in the mountains round Ben Nevis and Snowdon, in the English Lake District, and in the uplands of Cornwall and Devon. And it may fairly accurately be said that, as we cross England from west to east, the amount of the annual fall steadily decreases, as is shown in the accompanying map. The moisture-laden clouds, driven across the Atlantic by the prevalent S.W. winds, discharge their contents on meeting the cold high lands of western

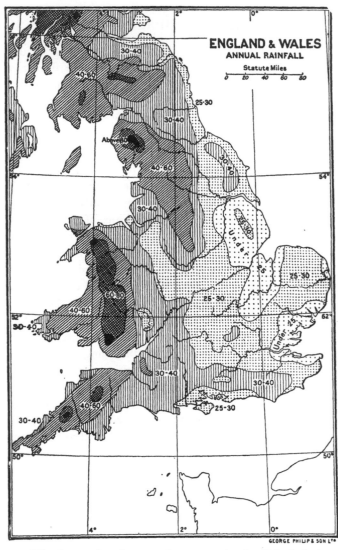

ENGLAND & WALES
ANNUAL RAINFALL

Statute Miles
0    20    40    60    80

*(The figures give the approximate annual rainfall in inches.)*

England, which thus act as a sort of umbrella, the driest parts of the country being on the east coast. While the average fall over a considerable part of the west side of it is from 40 to 60 inches, that on the east is no more than from 25 to 30; and round the Wash and in parts of Suffolk and Essex it is under 25 inches. Thus it sometimes happens that when the west of England has rain enough and to spare, the eastern districts are suffering from the want of it. The effect of this difference is shown in a marked degree in the character of the crops. The farms of the rainier west are to a great extent laid down in grass. The drier districts of the east grow more corn, which needs dry weather to ripen it.

The driest month in England generally is March, whose rainfall averages 1·46 inches; and the wettest is October, in which the average amount is 2·81 inches.

The pleasant climate of Devonshire, which is highly conducive to health and to extreme longevity, and which, especially in the south, favours luxuriant and even sub-tropical vegetation, owes its character to five main causes;—the fact that the county is bounded on two sides by the sea; the influence of the Gulf Stream or warm ocean current, which directly affects both coasts; the warmth and moisture of the prevailing winds; the shelter afforded to the southern districts by the high ground of Dartmoor; and the large amount of bright sunshine with which the county, and especially the south of it, is favoured.

It is owing to the Gulf Stream that the temperature of the sea in the English Channel is many degrees higher in winter, even as far east as the Goodwins, than that in the North Sea. Thus, the east wind, blowing over 200 miles of warmed water, has, by the time it reaches Devonshire, lost much of its proverbial bitterness. On the other hand it is mainly owing to the influence of the Gulf Stream that the climate of Ilfracombe is more equable than that of any other town in England except

Upcott Lane, Bideford

Falmouth, and is, in fact, from half a degree to a degree warmer in winter and cooler in summer than Torquay itself.

It is largely the shelter given to it by Dartmoor and the Blackdown Hills that makes the south and south-east of Devonshire so famous for its warm and pleasant climate. The south-west also benefits from the protection afforded by the spurs of Dartmoor, but the rainfall of that district is heavier and the air more relaxing.

The prevailing winds in Devonshire are the west and south-west, which, blowing across the open Atlantic, are also the chief rain-carrying winds. Both are comparatively warm, the latter following the course of the Gulf Stream. But they are often violent, and vegetation fully exposed to them does not flourish.

The peninsula of which Devonshire forms a part contains the warmest districts in Great Britain. The annual average temperature of the whole county is 49½° Fahr., which is a degree and a half higher than that of the whole of England; while the average for Torquay is 51°. In the three winter months, January, February, and March, in which the average temperature for London is 39·7°, that for Torquay is 41·3°.

The annual amount of bright sunshine in Devonshire naturally varies in different years. In 1906 it was nearly 2000 hours on the south coast, and not less than 1800 on the north. The average amount appears, however, to be about 1700 in the south, and 1500 on the north coast, or between three and four hours a day. The actual amount would, of course, be much more than this in summer, and much less in the winter.

The amount of rain in Devonshire in 1907, which was an average year throughout the country, was (taking the mean of the 169 stations named in Dr H. R. Mill's *British Rainfall*) 41·24 inches, falling on 210 days, or about 7½ inches and 7 days above the average for the whole of England and Wales. In the same year

196·16 inches of rain were registered at the Llyn Llydaw Copper Mine near Snowdon, and only 16·6 inches at Clacton-on-Sea. The wettest part of the county is Dartmoor, which catches the moisture-laden clouds coming up from the Atlantic. In 1907 some 81 inches of rain fell at Princetown, at a spot 1390 feet above the sea. This amount was, however, much exceeded in 1903, when the rainfall at the same station was 102·32 inches, and in Cowsic Valley, a little lower down, it was half an inch more. The driest part of Devonshire is the south-east coast. In 1907 only 26·27 inches of rain fell at Exmouth, for example. Heavy as the rainfall is, the slopes of the land are so steep and the soil in general so porous that the water soon runs away, with the result that both the earth and the air are drier than might be expected.

A feature of Dartmoor even more striking and characteristic than its heavy rainfall is the fog which so frequently covers it, and which is sometimes so dense as to cause the most experienced moor-men to lose their way.

Speaking generally, the climate of Devonshire may be described as warm and moist and remarkably equable. The winters are very mild, and snow is rare, except on Dartmoor. On the south coast of the county many plants which in less favoured parts of England need protection in the winter, such, for instance, as geraniums, hydrangeas, heliotropes, and camellias, are left out-of-doors all the year. Magnolias reach to the tops of the houses, myrtles grow to a height of thirty feet or more, palms and eucalyptus flourish, and oranges, lemons, and citrons do well in the open air.

Devonshire seems peculiarly liable to seismic disturbances, and many slight shocks of earthquake have been recorded.

## 12.    People—Race.    Dialects.    Settle=ments.    Population.

The earliest inhabitants of Devonshire, the people of the Palaeolithic or Early Stone Age, have left few traces beyond their weapons and implements of flint.    They lived in caves or on the banks of rivers.    They were hunters, and appear to have practised no craft but that of hunting, while their arts seem to have been almost if not entirely limited to the use of fire and to the making of rude instruments of stone.

But during the Neolithic Period, as the Later Stone Age is called, the district, it is believed, was invaded by an Iberian or Ivernian race from south-western Europe, a race possessing flocks and herds, with a knowledge of many arts and crafts, such as spinning and weaving, the making of pottery and of dug-out canoes, but having at first no acquaintance with the use of metal.    They were of the same stock as the Silures of South Wales, and were probably dark-haired and black-eyed, round-headed and short of stature.    Their descendants may, perhaps, still be seen in the county, especially on the skirts of Exmoor, and it is quite possible that their breeds of domestic animals may be represented upon Devonshire farms to-day.

K. D.                                                         7

The Iberians were, it is thought, conquered and driven westward by the very different Goidels or Gaels, a powerful Celtic race, tall, fair, long-headed, much further advanced in arts and crafts, and to some extent users of bronze for tools and ornaments. It is thought by some authorities that it was they who set up the stone circles, avenues, and menhirs, and who built the rude stone huts which still remain on Dartmoor. Many of the people of Scotland and Ireland are their descendants, and their language is still spoken in the Highlands and Western Isles of Scotland, in parts of Ireland, and in the Isle of Man.

The Gaels were, it is believed, succeeded and conquered in the fourth century before Christ by the Brythons, another Celtic race, who gave their name to our island. They took possession of Wales, and of Scotland as far as the Highlands, but they do not appear to have crossed into Ireland. They were, to a great extent, users of bronze, but they also worked in iron, and were the first of the Iron Age in this country. It is probable that they built most of the hill-forts of Devonshire, and that they made many of the roads, some of which were afterwards adapted and improved by the Romans.

Shortly before the landing of Julius Caesar, Britain was invaded by still another Celtic race, the Belgae from Gaul, a tall, dark-haired people, as may be gathered from the appearance of their descendants, the Walloons of Liège and the Ardennes.

The Roman tenure of Devonshire was of a very

limited character, and can have had little effect upon the inhabitants. The Saxon occupation of the district was more of the nature of colonisation than of conquest. By the time they had crossed Somerset the Saxons were, at least nominally, Christians ; and although their treatment of the original occupants was none too gentle, as, for instance, in the expulsion of the Britons from Exeter by Athelstan, it seems likely that the two races settled quietly down together, the Saxons probably becoming the land-owners, and the Celts the peasantry. It is thought that the main population of the county is Celtic, of one or other of the three waves of Celtic invasion. There are in Devonshire many Celtic place-names, especially of hills and rivers; and some of the latter, with the addition of Saxon endings, such as *ham*, *ton*, and *stock*, survive also in the names of towns. In some cases, curious English-looking names can be traced to Celtic words of quite another meaning. Thus, Bowerman's Nose, the name of a famous crag on Dartmoor, is probably a corruption of *Veor maen*, "the great stone."

The dialect of Devonshire, like the very similar speech of west Somerset, is Saxon, with strong traces of Celtic influence in its pronunciation. One of many peculiarities is the sound of the diphthongs *oo* and *ou*, which are pronounced like the French *u* or the German *ü*. Another peculiarity is the great variety of the vowel sounds, and the indistinctness or modification of some of the consonants. Again, *th* and even *v* are often sounded like *dh*. It is also very characteristic to put *d* for *th* as, for instance, *datch* for *thatch*, or *dishle* for *thistle*.

7—2

There are many words in common use in Devonshire which are almost or entirely unknown elsewhere, and which may be regarded as survivals of ancient Saxon or, in some cases, British speech. Such, for example, are :—

*whisht*, weird, uncanny, lonely, "overlooked."
*cloam*, earthenware.
*clam*, a foot-bridge.
*yark*, lively.
*stivery*, disordered.
*clunk*, to swallow.
*giglets*, young men or women seeking new situations.
*havage*, character.
*zamzoaky*, tepid.
*coochey*, clumsy.
*mawn*, a basket.
*frickety*, heavy, sodden.

*pluff*, not well.
*spraggetty*, spotted.
*dimpety*, dusk.
*chicket*, cheerful.
*spilsky*, lean.
*fess*, smart.
*plum*, soft.
*scamlin*, irregular.
*thurdle*, miserable.
*sklum*, to grasp roughly.
*yaw*, to bite.
*shugg*, shy.
*ippet*, a lizard.

It should be remembered that some of the forms of Devonshire dialect which strike the educated ear as ungrammatical are really survivals of pure Saxon speech, such as was in use at the courts of Alfred the Great and Athelstan. English in other parts of England has undergone great changes. In the West Country it has in some respects kept closer to the original forms.

The Norman Conquest left its mark in many places. Double names, such as Berry Pomeroy, Sampford Courtenay, and Wear Gifford, suggest the addition of a Norman family title to the existing Saxon name of a manor. It is quite possible that the common

Devonshire word *fay*, as in *Yes, fay* and *No, fay*, is a survival of the Old French *fay* (for *foi*), "faith."

Huguenots and other French refugees have also at

A Cockle Woman, River Exe

various times settled in the county, as, for instance, at Exeter, where they introduced the art of weaving tapestry, at Barnstaple, where they taught new and better methods of making cloth, and at Plymouth, where many took refuge after the Revocation of the Edict of Nantes. And it seems likely that Bratton Fleming and Stoke

A Honiton Lace-Worker

Fleming were named after Flemish immigrants, many of whom settled in Devonshire.

The population of the Geographical County of Devonshire, according to the census of 1901, is 661,314, and there are in the county 123,608 inhabited houses. A hundred years ago the population was 340,308; it has, therefore, not quite doubled during the century. In the busier county of Kent the population has in the same time increased from 268,097 to 1,348,841. In common with all except five of the counties of England, there are in Devonshire more women than men; the excess of the female over the male population being 37,096.

The chief occupation is agriculture, which provides employment for 42,000 people, or about one-nineteenth of the inhabitants—a considerably lower proportion than in the adjoining county of Somerset. Less than 3000 men are engaged in mines and quarries; 2000 are fishermen, and lace-making occupies 350 men and 1500 women.

In common with many other parts of England the small country parishes of Devonshire are much less populous than they were. In the last fifty years there has been a decline of 17,000 in the rural population.

Devonshire is a somewhat thinly-inhabited county. There are in it a little more than $2\frac{1}{2}$ acres to every man, woman, and child, or 254 persons to the square mile, compared with 558 to the square mile for the whole of England and Wales. Westmorland, the most sparsely-populated county, has only 82 people to the square mile, or eight acres to each inhabitant. Lancashire, on the other hand, contains more than 2300 people to the square mile, or four people to every acre;

and in the county of Middlesex there are 12,669 to the square mile, which gives about twenty inhabitants to every acre of ground.

## 13. Agriculture—Main Cultivations. Woodlands. Stock.

The area of all the land in England is, in round numbers, 32½ millions of acres, of which 24½ millions are under cultivation; 10¾ million acres being arable, and the greater part of the rest being devoted to permanent grass. For some years past the area of cultivation in the British Islands has been gradually growing less; and in 1908 the decrease in England alone was more than 25,000 acres, chiefly in the amount of land given up to barley and oats, but extending to almost all crops except wheat, potatoes, and lucerne, which showed a slight advance. The cultivation of fruit, especially of small fruit, continues to increase, but the total space devoted to it is not quite 300,000 acres.

With regard to live stock, the Government returns show that the total number of horses in England (about a million) was 10,000 less in 1908 than in 1907; but that the number of cattle (about five millions), of sheep (about sixteen millions), and of pigs (about two and a half millions) had increased, especially in the case of sheep and pigs.

Devonshire is eminently an agricultural county, having few industrial or manufacturing centres, and still fewer mining interests, although in the past it has been famous for weaving, and for tin and copper mining. There is in the county a great variety of soil, from almost barren sand to the rich alluvial earth of the many river valleys, such as the vales of Honiton and Exeter, for example, and that not very clearly defined tract of country called the South Hams, lying south of Dartmoor, including the

Old Ford Farm, Bideford

district between the Tamar and the Teign, and containing some of the most fertile land in England. The climate, as has been shown, is mild and equable, but the rainfall is heavy; and the farms of Devonshire, like those in the adjacent counties, are mainly devoted to pasturage, although fruit-growing is an important industry. Red Devon cattle are well known and highly valued; and the sturdy little ponies of Exmoor and Dartmoor have been famous since Saxon times.

According to the latest returns of the Ordnance Survey, Devonshire contains, exclusive of water, more than a million and a half (1,667,154) acres, of which nearly a million and a quarter (1,211,648) are under

Exmoor Ponies

cultivation, including rather more than 500,000 acres of arable land, and nearly 700,000 acres of permanent grass. The latter, which as will be seen is more than half the cultivated area, is more than twice that in Dorset or Cornwall, rather more than that in Somerset, and is only exceeded in the much larger county of Yorkshire. It may be added that the arable land was 11,000 acres less and the permanent grass 11,000 acres more in 1908 than in 1907.

Corn crops—which in the returns are made to include not only wheat, barley, oats, and rye, but peas and beans —occupy altogether about 200,000 acres, or one-sixth of the cultivated area. In this respect Devonshire surpasses the three adjoining counties, and is excelled by only six

Red Devon Cow

English shires; Essex and Lincoln, where corn crops occupy one-third of the area, Norfolk, where they are two-fifths, Cambridge and Suffolk, where they take up nearly one-half, and Yorkshire, where more than half the cultivated area is thus occupied. With regard to wheat alone, the average yield per acre in Devonshire, for the last ten years, is only 26¼ bushels, which is lower than that of any other county in England except Monmouth.

Green crops other than permanent grass, and roots, occupy altogether about 300,000 acres, an amount exceeded only in Norfolk and Yorkshire.

Devonshire ranks very high as a fruit-growing county, and the area of its apple-orchards, about 27,000 acres, was, in 1908, greater than that of any other county in England. Apples are grown in many districts, but especially in the Vale of Exeter, in the South Hams, and in the Valley of the Dart. Much of the fruit is, however, grown only for making cider, and is of little value for the table. Plympton is said to have had the first cider-orchard in England. When pears, plums, and cherries are included in the fruit returns, Devonshire takes third place, being surpassed by Kent and Hereford. Vines are grown against many cottage walls, as is the case in other southern counties; but it is remarkable, considering the mildness of the climate, that no Devonshire vineyard is mentioned in Domesday Book, although several are included in the survey for Somerset.

The space devoted to small fruits—strawberries, raspberries, currants, and gooseberries—although showing a large comparative increase over 1907, amounted in 1908 to no more than 1252 acres. In this respect Devonshire is fourteenth among the English counties, producing little more than one-twentieth as much small fruit as Kent, for instance.

Devonshire has no true forests. Dartmoor and Exmoor were so called in the sense of being unenclosed and uncultivated. But except on the moors, the county is well-timbered, and its fine trees add greatly to its beauty. Its woods, plantations, and coppices amount altogether to nearly 90,000 acres, or about one-eighteenth of its whole area; and it here ranks fifth among the shires of England. Sussex has the greatest proportion of woodland, about one-seventh of its total area; and Cambridgeshire, with only one-ninety-second, has the least. There are considerable woods in some of the many beautiful parks;

Gathering Cider Apples

but probably the most famous is the Wistman's Wood, near Two Bridges, the ancestors of whose stunted and fantastic-looking oak-trees are mentioned in Domesday Book.

The total number of agricultural holdings in Devonshire, in 1908, was nearly 15,000, or about one-twentieth of those in all England. This is greater than that of any other county except Lancashire, Lincoln, and Yorkshire. Nearly 3000 holdings are of five acres or less, and there are only six other counties which have more of these small farms.

The numbers of the various kinds of live stock in

A Water-mill at Uplyme

Devonshire are large, and the county ranks very high under the four main heads. In cattle (295,000, or 2000 less than in 1907) it stands second in all England, being surpassed only by Yorkshire; in horses (59,000, or 1500 less than the previous year) it is fourth; in sheep (900,000, or 29,000 more) it is fifth; and in pigs (106,000, or 5000 more) it is sixth. The average price per stone of fat Devon cattle was higher in 1908 than that of any other breed in England, and the value per head of three-year-old Devon store cattle was only exceeded by that of Herefords.

There is no cheese made in Devonshire to compare with the famous " Cheddar " of the neighbouring county; but Devonshire cream, although closely rivalled by that of both Cornwall and Somerset, is known all over the kingdom.

## 14.   Industries and Manufactures.

Devonshire, although in former ages famous even on the continent of Europe for its cloth-weaving, no longer ranks as a manufacturing county.   Apart from agriculture and fishing, its industries are now mainly confined to the making of lace and cider, to ship-building, and to the manufacture of earthenware.

The prevalence in the county of the names of Webber and Tucker is some evidence of the extent and antiquity of the woollen trade, which, from very early times, flourished all over Devonshire until the closing years of the eighteenth century, when it was greatly checked by the introduction of cotton fabrics.   One of the most important seats of the manufacture was Tiverton, where the industry was established in the fourteenth century, and reached perhaps its greatest height in the sixteenth.

It was his success as a cloth-merchant which enabled Peter Blundell to found here his famous school. The chief woollen market of the county was originally at Crediton, but it was removed in the sixteenth century to Exeter, which long ranked second only to Leeds, and in its palmy days exported annually more than 300,000 pieces of cloth. Other important centres of the trade were Barnstaple, where towards the end of the sixteenth century improved methods of weaving were introduced by French refugees; Tavistock, whose kerseymeres had a European reputation; Honiton, Cullompton, and Totnes. Now Ashburton and Buckfastleigh, where there are some manufactures of blankets and serges, are the only towns where the industry survives.

Lace-making, which has been a characteristic Devonshire industry for nearly three hundred years, is said to have been introduced at Honiton by Flemish refugees at the close of the sixteenth century, and to have been well established by 1630. The lace was a most costly product, chiefly because the special thread used in making it had to be imported from the Low Countries. In old days the price of Devonshire lace is said to have been reckoned by the number of shillings which would cover it. But the change of fashion in men's dress lessened the demand for lace ; and the introduction of machinery in 1808 greatly diminished its cost. The piece of lace which in the eighteenth century would have cost £15, could be purchased a few years later for 15s., and can now be obtained, machine-made, for 15d. There was some revival of the trade after the making, at Beer, of Queen Victoria's wedding-dress, at a cost of a thousand pounds. Schools were established for the training of lace-workers; and by

Devonshire Lace

1870 the industry provided employment for 8000 people. The manufacture has, however, again greatly declined,

K. D.                                                           8

and although there is a lace-factory at Tiverton, and although hand-made or pillow-lace is still worked in many cottages in the south-east, especially at Beer, Colyton, and Seaton, the total number of lace-workers in the whole county, at the last census, was less than 2000.

Carpets in imitation of those of Turkey were first made at Axminster in 1755, but in 1835 the looms were removed to Wilton, near Salisbury.

There are valuable deposits of various kinds of potter's clay in Devonshire, and although much of this is exported, a good deal is used in the county. There were formerly many small, scattered potteries in North Devon, but the chief seats of the industry now are at Bideford, where a good deal of rough pottery is made ; at Annery, noted for its glazed bricks and tiles ; and at Barnstaple, where are extensive and long-established potteries of what is called Barum ware, which has been compared to the Italian *sgraffito*. The potteries of Bovey Tracy, which use both local and imported clay, employ from 250 to 350 hands. The fine red clay of Watcombe is used to make terra-cotta ; and at Lee Moor, near Plympton, whence much kaolin or fine china clay is exported, the silicious refuse is made up into bricks of high quality for use in metallurgical furnaces. The kaolin deposits of Devonshire were discovered by Cookworthy, who made porcelain at Plymouth from 1772 to 1774, after which date the works were removed to Bristol.

Devonshire is one of the chief cider-producing counties, and its apple-orchards are the most extensive in our island. Some of the best varieties of apples for cider-making—an

Devonshire Pottery from the Watcombe Works

**Cider-making in the 17th Century**
*(From an old print)*

industry which is carried on throughout a very large part of the county though Totnes, Whimple, Crediton, Exeter

A Modern Cider Press

and Tiverton are perhaps the best known centres—are Kingston Black, both the sweet and the sour Woodbines (known locally as Slack-me-girdles), Sweet Alford and Fair Maid of Devon.

It is interesting to note that printing was early introduced into Devonshire.   In 1525 the fifth printing-press
in England was set up in Tavistock.   There are paper-
mills at Cullompton, iron-works near Kingsbridge, glove-
factories at Torrington, umber-works at Ashburton,
tanneries and shoe-factories at Crediton, and agricultural
implement works at Exeter.

Ship-building Yard, Brixham

In addition to the very important Government works
at Devonport and Keyham dockyards, there is a considerable amount of ship and boat-building, especially on
the Dart and at Brixham, and the industry employs
altogether about 3500 men.

## 15. Mines and Minerals.

There was a time when mining, especially tin-mining, was the most important industry of Devonshire. Traces left all over Dartmoor show that at a very early period tin was obtained there by the process called "streaming," that is to say by the washing of grains of the metal out of the disintegrated and crumbling granite. Vast numbers of abandoned shafts sunk in search of tin, copper, iron, manganese, and even silver, remain, together with their too often ugly buildings, as evidence of the former magnitude of the industry. At the present day, however, only twenty-four mines are in active operation, providing employment for no more than 700 men, who, in 1907, raised less than 1700 tons of metal of all descriptions.

The tin-miners of Devon and Cornwall were early formed into a corporate body whose affairs were managed by a Stannary parliament that met on Hingston Down. At a later period, probably at the beginning of the fourteenth century, the Devonshire men held their own parliament, which assembled on Crockern Tor. They were governed by a Warden—Sir Walter Ralegh held the office for some years—appointed by the Duchy of Cornwall, who collected the Duchy dues or royalties, and having ascertained the purity of each block of tin by "coinage," that is, by cutting off for analysis a " coin " or corner, stamped it with the Duchy arms. A miner convicted of selling impure tin was punished by having some of the melted metal poured down his throat.

Lydford, Tavistock, Chagford, and Ashburton were called Stannary Towns, since at each of those places blocks of tin might be tested; and the arbitrary nature of the Stannary Court is hinted at in the proverbial expression : " Lydford Law ; hang first and try afterwards."

The chief mining district of Devonshire begins at the Tamar and extends across Dartmoor and along its borders. The most important centre was Tavistock, but there were rich mines at North Molton, where several kinds of metal were worked, near Ashburton, and elsewhere.

The principal ores are those of copper and tin ; some iron is worked, and there are rich veins, believed to be not yet exhausted although not now worked, of galena or silver-lead.

The Devonshire mines formerly produced more tin than those of Cornwall; but since the fourteenth century the output of the latter county has been the greater. The quantity now raised in Devonshire is inconsiderable, and in 1907 amounted to only 94 tons. Tin is very largely used in making what is called tin-plate, which is really sheet-iron dipped into the melted metal. It is also mixed with copper to make bronze or machine-brass, and it was for the manufacture of bronze that it was so much sought after by the ancient inhabitants of the country.

Copper-mining in Devonshire is believed to be a comparatively modern industry. It is not known whether the ancient Britons made their own bronze from native tin and copper, or whether they imported it from abroad. The Devon Great Consols Mine, four miles from Tavistock, once by far the richest mine in England, and one of the richest in the world, has shipped as much as 1200 tons of copper in a single month, and has produced altogether 3½ million pounds' worth of ore ; but it now yields little but arsenic.   In 1907 only 652 tons were raised in the whole county.   The ore is not exhausted, but the cost of raising it from deep mines is too great to withstand foreign competition.

Devon Great Consols Mine

Ores of iron and zinc are widely distributed, but are little worked ; and the annual yield, both of these metals and of manganese, of which this county was once a chief source of supply, is inconsiderable.   Very rich silver-lead ore was formerly worked at Bere Alston and at Combe Martin, but the mines in both places have been abandoned. A very massive cup, made of Combe Martin silver, given to the Corporation of London by Queen Elizabeth, is still used at the inauguration of each Lord Mayor. Arsenic and arsenical pyrites, ochre, and umber are obtained, especially from some mines whose more valuable ore is exhausted. Cobalt, tungsten, and uranium also occur, and gold has been found in small quantities, generally in streams, as, for instance, in the West Webburn and below Lethitor.

Although the metal mines of Devonshire have lost their old importance, there are other minerals of great commercial value, of which altogether more than a million tons are obtained in the course of a year. China clay, or kaolin, a product of the natural decomposition of granite, is worked at Lee Moor, and more than 75,000 tons—which, however, is only one-tenth of that obtained from Cornwall—are annually exported, especially to Staffordshire, for the making of fine earthenware. Other kinds of potter's clay, white at Kingsteignton and Bovey Tracy, and red at Watcombe, are dug in still larger quantities.

There are many quarries in Devonshire, the most important of them being of limestone, of which more than half a million tons are worked every year. Heytor granite was used in London Bridge and Waterloo Bridge, and Lundy granite in the Thames Embankment. But the stone is not considered equal to that from Cornwall. The same remark applies to the slate, of which only 5000 tons are now raised annually. There are old quarries of it near Kingsbridge, and also at Tavistock and other places. Colyton slate is used for billiard-tables.

Marble is worked at Chudleigh, and a finer quality at Ipplepen, Torquay, and Plymouth. There are large quarries at Beer. The material for whetstones has long been dug in the Blackdown Hills, where the refuse from the workings, like lines of railway embankment, is a feature in the landscape.

Stone Quarry, Beer

There is no coal in Devonshire, but there is much lignite at Bovey Tracy, where, in the bed of an ancient lake, a deposit of layers of it occurs, alternating with clay and sand, to a depth of 100 feet. On account of its disagreeable smell while burning and its low heating-power it is not used for fuel except for firing bricks, and to some extent in the pottery-kilns. There are also extensive beds of anthracite or culm near Bideford ; but this, again, is not of a quality to serve as fuel, except for lime-burning, and the product of the one solitary working is ground up to make a paint called Bideford Black. There are vast and valuable deposits of peat on Dartmoor, in some places as much as thirty feet deep.


## 16.　Fisheries and Fishing Stations.

The fisheries of the British Islands form one of our most important industries, providing regular or occasional employment for nearly 100,000 men and boys in the catching of the fish ; for a very great number of persons engaged in secondary occupations connected with the industry, who probably far outnumber the actual fishermen; and for innumerable people of all grades engaged in distributing the eight million pounds' worth of fish brought into the ports of England and Wales each year by British ships alone. The fisheries also furnish an immense quantity of cheap and wholesome food, which, by rapid methods of transit, is available in all parts of the country.

By far the most productive of our fishing-grounds, although not as predominant as it was some years ago, is the North Sea—an area of more than 150,000 square miles, in which are taken more than half of all the British-caught fish, not including shell-fish, which are annually landed on the coasts of England and Wales. More fish are brought, every year, into Grimsby, Hull, Lowestoft, and Yarmouth than into all the other fishing-ports of

England put together. It is interesting to note that, while according to the latest returns there were 1731 British steam-trawlers and drifters, exclusive of ordinary fishing-boats, engaged in the North Sea fisheries, there were only 451 similar craft belonging to the ports of Germany, Holland, Belgium, and France put together. In other words there are four British steam-trawlers in the North Sea to every foreigner. Much fishing is also done by English trawlers off the shores of Iceland, Norway, and the Faroës, and the boats now go as far even as the White Sea and the coast of Morocco.

About half the fish are taken by trawling, which consists in dragging a beam of wood, with a net attached to it, along the bottom of the sea, in comparatively shallow water. Very many different species are caught in this way, but haddock, plaice, and cod are by far the most numerous, and make up between them nearly half the total amount of all the fish landed in England and Wales in a year. Much fishing is also done with seine nets, or with drift nets, both of which are long nets, attached to floats of cork or to air-bladders and let down into the sea without regard to the depth, and sometimes at a considerable distance from the shore. Stake nets, fastened to poles fixed in shallow water near the land, are also much used. Herrings are the chief fish caught in drift nets and seines, and more of them are landed than of any other kind of fish. The latest return gives the total quantity of herrings annually brought into English ports as rather more than 200,000 tons, of cod as about 100,000 tons, and of plaice as about 50,000 tons. Pilchards, which are full-grown sardines, and much resemble herrings in appearance, are caught in large quantities—which, however, seem trifling in comparison with those of the three fish named above—in seine nets off the coasts of Cornwall and Devon, and nowhere else in the British Isles. Many fish, especially halibut, cod, and ling are taken with hook and line, sometimes at great depths. Crabs and lobsters are caught in

Fish Market at Brixham

wicker traps or baskets called pots, and oysters are usually taken by dredging.

In spite of its two long stretches of seaboard, the fisheries of Devonshire are not equal in productiveness to those of Cornwall, and are insignificant in comparison with those of the east coast; and the total value of the fish landed at all its ports taken together amounted, according to the latest return, to no more than £150,000, or one-nineteenth of what is landed at Grimsby alone. It was far exceeded at five other east-coast fishing stations.

The fish of the English Channel differ considerably from those of the North Sea. Haddock, the most abundant species on the east coast, is very rare in the south, and practically none are caught at any of the Devonshire fishing stations. The cod, again, is a northern species, and is almost entirely absent from both the Bristol and the English Channels. Whiting is one of the most abundant of English Channel fish ; and in this species, as well as in soles and turbot, the south coast is of all the British fishing-grounds second only in productiveness to the east coast. More conger-eels are caught in the English Channel than anywhere else off our islands, and there is also a great abundance of gurnards, skates, and dogfish.

The Devonshire fishermen catch great quantities of whiting, herring, mackerel, sprats, and pilchards, together with considerable numbers of soles, turbot, plaice, pollack, skates, congers, crabs, lobsters, and prawns. Herrings were formerly very abundant off Lynmouth. The last great shoals appeared in 1823. A skate caught off the south coast of Devonshire measured nine feet by six and a half feet, and weighed 560 pounds. The quantity of sprats annually caught in Devonshire waters is very great, but, as in other districts, varies very much in different years. Thus the amount brought into Torquay in 1905 was more than 500 tons, or more than were landed at any other port in the kingdom ; but in 1906 the quantity was only 100 tons. Pilchards, as has been already observed, are confined to Cornwall and to the south coast of Devon ; but by far the greater quantity are taken at the fishing stations of the former county. Almost all the pilchards caught in Devonshire waters are landed at Plymouth. None are taken further east than Dawlish. These fish, which are particularly oily, are mostly salted and exported to the Mediterranean. Dog-

Brixham Trawlers

fish, which are very abundant and formerly thrown away as worthless, are finding an increasing market, especially in London, where they are filleted and sold as " flake."

The most important fishing stations are Plymouth, Brixham, and Torquay, the annual value of whose fisheries according to the latest return is about £66,000, £60,000 and £8,000 respectively. There is also a good deal of fishing off Exmouth, Teignmouth, Dartmouth, Torcross, and Budleigh Salterton, where the annual values vary from £4000 to £750 a year. It is interesting to compare these figures with the annual value of the fish brought into Grimsby, which, by the last return, amounted to nearly three millions sterling.

There are valuable salmon fisheries at Exmouth, Teignmouth, and Babbacombe; and most of the Devonshire streams abound with trout, although the fish as a rule run small. Thirteen Devonshire fisheries are named in Domesday Book. The most valuable was that at Dartington, for which two fishermen paid a yearly rent of eighty salmon.

## 17. Shipping and Trade.

The ports of Devonshire once ranked among the first in England, and her sailors have for many centuries been famous for their enterprise and daring. It was from this county that the first English trading-expeditions sailed to Africa, Brazil, and North America. They were Devonshire men, who, by taking possession of Newfoundland, established the first English colony— in which most of the old families are of Devonshire descent. Devonshire ships were long the terror of the Spanish Main. Devonshire men were among the very foremost in the defeat of the Spanish Armada. A Devonshire captain was the first Englishman to sail round the world; and although we remember with regret that his friend and comrade was the first Englishman to

K. D.                                                              9

engage in the iniquitous traffic of the slave-trade, we are proud to think that few men did more than he to improve our ships and the condition of our seamen.

In their palmy days, in the century or more following the flight of the Armada, Bideford and Topsham had each of them more trade with the young colonies of North America than any other English town except London.  Barnstaple, Ilfracombe, Dartmouth, Brixham, and Appledore were once important seaports.  At the present day not one of the whole seven has sufficient trade to be honoured with a separate entry in the Government Shipping Returns.  Plymouth is now the only maritime town of commercial importance.  Even its traffic, large as it seems, is small in comparison with that of London or Liverpool, and as far as trading statistics go, it stands no higher than thirtieth among the ports of the United Kingdom.

Several causes have contributed to the decay of the Devonshire ports.  Most of them are situated on river-estuaries which, in the lapse of ages, have become silted up by mud and sand brought down by the rivers, or obstructed by shingle washed up by the waves.  The harbour of Sidmouth was destroyed by the encroachment of the sea and the fall of the cliffs which formerly protected it.  Again, the tonnage of ships, and consequently the amount of water they draw, have very greatly increased since Tudor and Stuart times, when these ports were in their prime; and it would be impossible for the large vessels of to-day to navigate the shallow and danger-strewn waters of our estuaries, even if they could cross the bars by which they are obstructed. Nor is it worth while to improve the navigation by dredging, as is done to so great an exent on the Thames, the Mersey, and the Clyde. The industries of Devonshire are now of small importance, and the county has no great manufacturing centres to supply freights. Plymouth Sound is the only busy waterway, and Plymouth is the one populous town requiring large quantities of imports.

The only harbour in North Devon given in the shipping returns is Barnstaple, with which are associated Ilfracombe, Bideford, and Appledore. Ilfracombe, the only port in the long stretch of coast between Bridgwater and Padstow, had formerly a good deal of traffic with Wales and Ireland, but its tidal harbour is now visited only by excursion steamers and small coasting-vessels.

The other three towns are river-ports. Barnstaple is eight miles from the mouth of the Torridge, Appledore is just inside the entrance of the Taw, and Bideford is five miles up the same river, whose estuary is obstructed by a dangerous bar, only to be crossed at high tide. In the sixteenth and seventeenth centuries all three towns had an active trade with North America; and even in comparatively recent times ships of 1000 tons have been moored at Barnstaple quay. But at the present day only 37 cargo-carrying vessels sail to and from the whole group in a year, and the trade (chiefly in timber and dye-stuffs) of all four together, principally with home ports, but also with Sweden and Norway and some other European countries, amounts to little more than £18,000 in twelve months.

Dartmouth, Brixham, and Salcombe form another group of ports, all of which have played a part in history. Their total trade, the import of timber and the export of ships and boats, amounts to nearly £19,000 a year, and such foreign intercourse as they have is chiefly with Sweden, Norway, and Russia.

Exeter and Exmouth, with which is associated Lyme Regis in the adjoining county of Dorsetshire, rank next in importance, Topsham, the ancient port of Exeter, having gone entirely to decay. Their annual trade, about half of which consists of wood, cured fish, and sugar, and which, as regards foreign intercourse, is mainly with France, Germany, and Sweden, is nearly £100,000.

The trade of Teignmouth and Torquay together, in the importation of paper-making materials and timber, and in the export of China-clay, chiefly with English ports, but also with France, Germany, Belgium, and Norway, amounts to £110,000 in the twelve-month.

Plymouth which, as has been already pointed out, is the only large sea-port in the county, has four times as much trade, and is entered and cleared by four times as many vessels as all the other ports of Devonshire put together. Its chief imports are grain (£540,000), timber (£250,000), sugar (£135,000), guano and manures (£110,000), and petroleum (£52,000); and its principal export is £52,000 worth of clay. Its imports and exports taken together amount to 1¾ million pounds sterling, and it is entered and cleared by 1656 ships in a year. Its chief foreign trade is with France, but its commerce may

Teignmouth

truly be said to be world-wide. Thirteen lines of ocean steamers sail from or call at Plymouth, the principal of which are the White Star, American, Norddeutscher-Lloyd, and Hamburg-American for the United States; the Orient and Peninsular and Oriental for Australia; the latter and the British India for India; the Shaw-Savill and New Zealand SS. Co. for New Zealand; and the British and African for the West Coast of Africa. There are also regular sailings of steamers for France, Scotland, Ireland, the Channel Isles, and various home ports.

It is interesting to compare this sea-traffic with that of London, which is entered and cleared by 18,491 cargo-carrying ships in the course of twelve months, and has a total annual import and export trade of 333 millions of pounds sterling; which are respectively about 11 times, and about 190 times as large as the corresponding figures for Plymouth.

But although Plymouth is a place of considerable maritime trade, a busy fishing station and a port of call for ocean-going steamers, for whose accommodation are provided spacious docks and ample quays, its greatest importance and renown—remembering that we include with it its sister towns of Stonehouse and Devonport—rest upon its rank as a naval station, as an arsenal which is second only to that of Woolwich, and as a naval dock-yard which is the largest in the kingdom.

The anchorage at the head of the Sound, once very much exposed and dangerous in southerly winds, is now protected by a stone breakwater nearly a mile long, designed to shelter ships of the Royal Navy. It was

Drake's Island from Mt. Edgcumbe, Plymouth

commenced in 1812 by Rennie, continued by his son, modified in the slope of its sides and improved in stability by violent storms, especially in 1817, and completed in 1840.

Taking all its various features into account, its commerce, its passenger traffic by means of ocean-liners and other steamers, its fisheries, its docks and dockyards, its barracks, its factories of marine appliances, its arsenal, and lastly the vast number of ships of all sizes, belonging to the navy, to the mercantile marine, or to the fishing-fleet, that are constantly leaving or entering the Sound, Plymouth is one of the most important sea-ports in the British Empire.

## 18. History of Devonshire.

The history of our country begins with the Roman occupation. For although we have ample and striking traces, in the shape of earthworks and stone circles, tools and weapons, pottery and ornaments, of the successive races of men who lived here before Julius Caesar set foot in Britain, those ancient and primitive people left no written records, not so much as an inscription on a single coin, and our knowledge of them is in the highest degree vague and uncertain.

Of many parts of our island the Romans took complete possession, constructing fortresses, making roads, establishing towns, building baths and temples and

luxuriously appointed villas, and scattering, wherever
they went, the coins whose lettering and devices have
revealed to us so much concerning the some time masters
of the world. In Somerset, for example, to which the
conquerors were attracted partly by the hot and health-
restoring springs of Bath, and partly by the silver-bearing
lead mines of the Mendip Hills, the relics of their
occupation have been found from one end of the county
to the other.

In Devonshire, on the other hand, such relics are so
few, and are confined to so limited an area that we are
driven to the conclusion that, except as regards the city
of Exeter, there was no definite Roman occupation at all.
There is probably not one camp of Roman workmanship
in the whole county. It is doubtful if any Roman road
went farther than the river Teign. The sites of only
two Roman villas are known with certainty. And
although Roman coins have been found in many places,
sometimes in hoards of hundreds, and in one case even
of thousands, they are not absolute proof of actual
occupation. The names Chester Moor, Scrobchester,
and Wickchester, all near the Cornish border, may,
perhaps, be of Roman origin.

There is, however, no doubt that Exeter, believed to
be the *Isca Dumniorum* of Antonine's *Itinerary*—that
wonderful register, planned by Julius Caesar and carried
out by Augustus, of distances and stations along all the
roads in the Empire—was an important Roman town ;
and there is reason to think, from the coins that have
been found at many points within the walls, that the city

was held by the Romans from the latter half of the first century of the Christian era until the time when the legions were recalled from Britain. The site of *Moridunum*, the second Roman station mentioned in the *Itinerary*, has not been identified, but there is some ground for the theory that it was at Hembury, four miles from Honiton.

A few vague and brief allusions in the Anglo-Saxon Chronicle, believed to refer to this county, describing how "Ina fought against Geraint," how "Cynewulf fought very many battles against the Welsh," and how "Egbert laid waste West Wales from eastward to westward," contain practically all that we know of the Saxon conquest of Devonshire. There is, indeed, so little record of actual fighting that it seems probable that the invaders settled here rather as colonists than conquerors, although Athelstan appears to have found it necessary to expel from Exeter the Britons who had so far shared the town with the Saxons.

The chief events in Devonshire between the departure of the Romans and the Norman Conquest were the repeated descents, spread over a long period of years, of the pirates whom we speak of as Danes or Northmen or Vikings; who pillaged the coast towns, sacked Exeter, sailed up the Tamar, and burnt and plundered Tavistock and Lydford. Victory was not always on the side of the marauders. Their first raid, in 851, was repulsed with great slaughter; and when, five and twenty years later, Guthrum seized Exeter, King Alfred promptly drove him out of it.

During the Saxon period there were mints at Exeter, Barnstaple, Totnes, and Lydford, and thousands of Devonshire-struck silver pennies are in existence. By far the greater number of them are in the royal museum at Stockholm, the most numerous being those of Ethelred II and Canute. Of the former there are in Stockholm 2254 specimens, compared with 144 in the British Museum. These Swedish specimens probably represent partly the plunder carried off by the Northmen, partly the bribes vainly paid to the invaders by Ethelred (whose surname of Unradig, " he who will not take counsel," or "the

Penny of Ethelred II, struck at Exeter

headstrong," has been misrendered " the Unready "), and partly the results of commerce while Canute was king.

The year succeeding the Battle of Hastings found William the Conqueror before the gates of Exeter, a place already regarded, as it continued to be for many centuries, as the key of the West of England. He took the city after a brief siege and proceeded to secure his hold upon it by building the castle of Rougemont, which was hardly finished when it was unsuccessfully attacked by the Saxons. A year later the sons of Harold also tried in vain to take it. The last man of mark in

Devonshire to hold out against Norman rule was Sithric, the Saxon abbot of Tavistock, who, when all was lost, fled to Hereward's camp of refuge in the Fens. A few Englishmen were left by the Conqueror in possession of their estates; but the county, as a whole, was divided among a number of Norman nobles, some of whose descendants, Courtenay, Carew, and Champernowne, for example, still survive in Devonshire. An interesting link with Norman times and customs is the ringing of the curfew bell, which is still kept up at Exeter, Okehampton, and other places. At eight o'clock every evening thirty strokes are sounded for "Curfew," and then eight more for the hour.

In the stormy reign of King Stephen Exeter was the last place to hold out for Queen Maud. The king was admitted into the town by the citizens, but the castle of Rougemont cost him a three months' siege.

The importance of Devonshire sea-ports brought the county into great prominence in mediaeval times. Part of Richard Cœur-de-Lion's crusading fleet, we are told, assembled at Dartmouth—a town which Chaucer, probably regarding it as a typical sea-port, chose for the native place of the Shipman in the *Canterbury Tales*. No other part of England furnished so many ships and men for Edward III's expedition against Calais. Again and again, in the fourteenth and fifteenth centuries, the French, in reprisal for what they had suffered by the attacks of England, harried the coast of Devon, plundering and burning Teignmouth, Plymouth, and other places on the coast.

The Black Death, the most terrible and destructive epidemic of which we have any record, which devastated the whole of England in 1348 and 1349, was very severe in this county, paralyzing agriculture and trade, and stopping for a time the building of Exeter cathedral.

Fighting in Devonshire during the Wars of the Roses was confined to an unsuccessful and half-hearted siege of Exeter by the Yorkists, and to attacks on the fortified manor-houses of Shute and Upcott. But many men in the county took sides in the struggle, and some of the great families suffered severely. Sir William Bonville was beheaded after the second Battle of St Albans. Of the ancient house of Courtenay, Thomas, Earl of Devon, was executed at York, Sir Hugh was beheaded at Sarum, and Sir John was killed at Wakefield Green. The county as a whole was Lancastrian. Queen Margaret herself was there after her defeat at Barnet; and French gold coins found in Blackpool sands are believed to be relics of the landing there, in 1470, of Warwick and Clarence. But when in the same year Edward IV visited Exeter, he was so well satisfied with his reception that he presented the corporation with a sword of state, which is still carried in processions before the mayor.

The peace of Devonshire in the fifteenth century was further disturbed by a rising, in 1483, against Richard III; by the march through the county, in 1497, of an army of Cornishmen who had risen in revolt against a heavy war-tax, and who were ultimately beaten at Blackheath; and by the insurrection, also in 1497, of

Perkin Warbeck, who claimed to be the Richard Duke of York usually said to have been murdered in the Tower, and who made a desperate although vain attack upon Exeter, when some of his men fought their way into the town, but were driven out again by the citizens.

About fifty years later, in 1549, there was a widespread and determined and altogether much more serious rebellion called the " Commotion," caused partly by the suppression of the Monasteries, which was greatly objected to by the poor, and partly by the introduction of the Prayer Book. The insurgents, who had collected from all parts of the West Country, and who were led by such men of mark as Pomeroy, Arundel, and Coffin, laid siege to Exeter and Plymouth, and for a time held the king's troops helplessly at bay. In the end, however, Lord Russell, one of the newly-created Lords Lieutenant, aided by German cavalry and Italian arquebusiers, defeated the rebels with great slaughter in a series of hotly-contested battles. The vicar of St Thomas in Exeter, who had encouraged the rising and who was described as very skilled both with the long-bow and the hand-gun, was hanged " in his Popish apparel " on the tower of his own church, and his body was left there for four years.

The reign of Queen Elizabeth has been called the Golden Age of English History. And among the heroic figures of that stirring time there are few more striking than the little group of Devonshire men who played so gallant a part in making England great:—Drake and Hawkyns, the scourges of Spain; Ralegh, courtier and soldier, sailor and author ; Gilbert, the discoverer of Newfoundland ; and Grenville, who at Flores, in the *Revenge* of immortal memory, kept at bay a Spanish fleet of fifty-three sail.

The great event of the reign was the defeat of the Armada. And although Howard of Effingham, the Lord High Admiral, showed himself a skilful and intrepid sailor, it is Drake whom we always think of first in

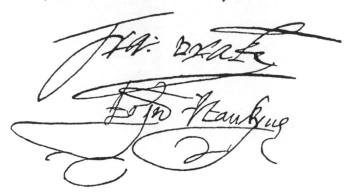

Signatures of Drake and Hawkyns

connection with the victory. It was Drake whose buccaneering exploits on the coasts of Spain and her colonies did so much to heighten Philip II's ambition to humiliate England. It was Drake who, in 1587, dashed into Cadiz, where the Armada was preparing, and by destroying 100 ships and vast quantities of stores, delayed for a whole year the sailing of the expedition. And when at last the Armada had been sighted, it was Drake who, according to the commonly received tradition, thinking it wiser to wait until the enemy's fleet had passed Plymouth Sound, and so take them in the rear, persuaded his fellow-captains to stay and finish that

Flagon taken by Drake from the "Capitana" of the Armada
(*In Windsor Castle*)

never-to-be-forgotten game of bowls, before the English ships, lying ready in the Catwater, should slip their moorings and stand out to sea. It was Drake who was foremost in the attack. It was Drake who took the *Capitana*, the flagship of Pedro de Valdez, and brought her, the first prize of the great victory, into Torbay. And when the English fireships had scattered the hostile fleet in headlong flight, it was Drake who was foremost in the chase.

Drake's Drum

Among the relics of Devonshire's greatest hero, carefully treasured by his descendants at his old home at Buckland Abbey, are his sword and the famous drum that he carried with him round the world; while at Nutwell Court are flags that he flew while in command of the *Pelican*, the miniature of herself given to him by Queen Elizabeth, and other objects of the greatest historical interest. And among the Royal plate in Windsor Castle is preserved the noble wine-flagon of bold silver-gilt *repoussé* work, standing nearly a yard high, which Drake took from the *Capitana* of the Armada and presented to Queen Elizabeth. An illustration of it, from a photograph taken for this book by command of the Queen, is here shown for the first time.

The "Mayflower" Stone on Plymouth Quay

During the retreat to Spain a second Armada vessel, the hospital-ship *St Peter the Great*, was driven ashore in Hope Cove; and the pulpit of St James's church, Exeter, and the timber roof of Tiverton School were, it is believed, made of wood either from this ship or from the *Capitana*.

It was not until this period that Plymouth came into prominence as a naval station. A special tax was levied on the pilchard fishery to provide money for the fortifi-

cations, and a leat or water-course was constructed with
the primary object, it is said, of supplying fresh water for
the royal ships.

A memorable event in James I's reign was the sailing
of the *Mayflower*. Preceded by another ship called the
*Speedwell* she set sail from Leyden in the autumn of 1620,
having on board a number of Puritan refugees bent on
finding in North America the religious freedom denied
to them in England. The two vessels having met at
Southampton and put into Dartmouth, were finally driven
back by stress of weather into Plymouth, whence—her
consort having proved unseaworthy—the *Mayflower* alone
continued the voyage, ultimately landing her 101 exiles at
Plymouth, Massachusetts, which, however, had received
its name five years before.

In the Civil War between Charles I and the Parliament, few counties saw more fighting than Devonshire.
The fighting consisted, however, not of pitched battles,
but of sieges and attacks on fortified positions; which,
indeed, was characteristic of the whole war, in whatever
part of the country it was waged. Every Devonshire
town of importance, a great number of villages, many
castles, manor-houses, and even churches played a part
in the struggle.

As a whole, the towns, with the exception of Exeter,
sympathised with the Parliament, while the rural districts,
encouraged by the great landowners, were mainly for the
king. The royal forces were, however, numerous in
Devonshire; Goring's army, in 1642, was 6000 strong;
and although fortune wavered, and although towns were

taken and retaken, there came a time, before the arrival of Fairfax and the New Model Army, when the royal standard flew from nearly every important town in Devon, Somerset, and Cornwall. There was one conspicuous exception. The party of the people never lost its hold on Plymouth, which, at a cost of 8000 lives, or more than the entire population of the town, withstood a blockade lasting from 1642 to 1646, together with many desperate attacks by Hopton, Prince Maurice, and the King himself, enduring altogether a longer siege than any other town in England.

Exeter was early seized for the Parliament, but the majority of the citizens were Royalists, and the city, which was regarded as one of the strongest Cavalier holds in the west, was soon retaken by Prince Maurice. Queen Henrietta was there in 1644, and there King Charles's youngest daughter, afterwards Duchess of Orleans, was born. When Fairfax retook the town in 1646 he allowed the garrison to march out with all the honours of war.

That year saw the final ruin of the Royal cause in the west, and the dispersal of the only army which, although little better than a mob, still kept the field for the King.

One of the most important, and at the same time most fiercely-contested Parliamentary victories, was the storming of the town of Torrington by Fairfax, at midnight, in the winter of 1646. After the battle, the church, which had been used by the king's troops (as also was Exeter cathedral) as a powder-magazine, was blown up, and 200 Royalist prisoners who had been confined in it and many of their guards were killed. The loss of Torrington was the death-blow of the Royal cause in Devon. All that a brave man could do, Hopton did. But the county was sick of the Royalists and their methods. The people had learnt that the well-disciplined troops of Fairfax were not mere robbers, like the ruffians of Grenville and Goring; and after Torrington the Royal army melted away.

The last place in the county to hold out for the King except Lundy, where there was no fighting, but which did not surrender until 1647, was Clifton Castle, or Fort Charles, near Salcombe. After enduring a blockade and siege of four months, with the trifling loss of one man killed and one wounded, the besieged were granted the same terms as the garrisons of Exeter and Barnstaple, and marched out with matches lighted, drums beating, and colours flying.

After the Battle of Worcester, in 1651, Charles II took refuge for a time in Devonshire. Four years later there was an attempt at an insurrection in his favour, known as Penruddock's Rising, and Charles was proclaimed King at South Molton. The movement was promptly suppressed, and its leader, Colonel Penruddock, was executed. It is interesting to remember that, at the Restoration in 1660, Exeter was the first town in England to acknowledge Charles II, and that he was there proclaimed King ten days before he landed at Dover. Seven years later, in 1667, the great Dutch Admiral De Ruyter captured all the shipping in Torbay.

During the Commonwealth and later, when there was no copper coinage in this country, many tradesmen all over England struck money of their own, chiefly in the form of farthings. Nearly 400 varieties of Devonshire "tokens," as they were called, issued by sixty different towns, are known. Ninety-one were struck at Exeter alone, which is more than were issued from any other provincial town except Norwich. At a much later period shillings and sixpenny tokens of leather were in circulation at Hartland.

After the Duke of Monmouth landed at Lyme in Dorsetshire, in 1685, Axminster was the first town that he occupied, and a number of Colyton men are said to have joined his army. Otherwise the rebellion hardly touched Devonshire. Yet Judge Jeffreys put to death, at various places in the county, thirty-seven of the Duke's misguided followers. After the Battle of Sedgemoor, Wade and other fugitives attempted to escape by sea from Ilfracombe, but they were obliged to put back, and were caught in the woods near Lynton.

When, under very different auspices, William of Orange landed at Brixham on the 5th of November, 1688, he marched to Exeter, as the chief city of the west. The citizens at first held aloof, but in the end they gave to the Deliverer their hearty and most valuable support.

Two years later the French admiral Tourville, fresh from his victory over our fleet off Beachy Head, landed a strong force at Teignmouth, and sacked and burnt that part of the town which ever since has been known as French Street.

Both in 1715 and in 1745 the county was suspected of showing sympathy with the exiled Stuarts. But when in the former year the Duke of Ormond, with a small party of French soldiers, appeared in a war-ship off Brixham, expecting to be welcomed by the Devonshire Jacobites, he met with no encouragement.

Several episodes in the history of Devonshire are associated with the French war of the close of the eighteenth and the beginning of the nineteenth centuries. In 1797 a French squadron, afterwards captured by Admiral Bridport, sank the fishing-boats in Ilfracombe harbour. A few years later, in 1809, Princetown was built for the reception of the rougher class of French prisoners of war, and during the five years that followed 12,679 Frenchmen and Americans were confined there, while many others were billeted at Okehampton, Ashburton, Tavistock, and Moreton Hampstead. The Princetown buildings were afterwards disused—except for a short occupation by a naphtha company—until 1850, when they were converted into a prison for convicts.

It was at this period that Torquay came into note, having become a place of residence for the families of naval officers serving on men-of-war anchored in Torbay. In Torbay, too, was enacted what may be called the last scene of the war. For it was here, on the 24th of July, 1815, that Napoleon was brought a prisoner. And here he remained, except for a few days spent in Plymouth Sound, until the 11th of August, when he was taken to St Helena.

## 19. Antiquities.

To the archaeologist and the antiquarian Devonshire is one of the most interesting counties in England. With the exception of Cornwall no other district is so rich in

Palaeolithic Flint Implement
(*From Kent's Cavern*)

relics of the ancient inhabitants of Britain; and it was from the caves of this county that Pengelly obtained that clear evidence of the extreme antiquity of man in this island which proved that he lived here, unnumbered ages back, when such animals as the cave-lion, the hyaena, and the mammoth ran wild in what we now call England.

The people who inhabited Britain before the coming of the Romans are said to have belonged to the Stone Age, the Bronze Age, or the Iron Age, according to the material which they used for their tools and weapons; and the first of these epochs is further divided into the Earlier and the Later Stone Ages, or the Palaeolithic and the Neolithic Periods. The implements of the former were very roughly fashioned of chipped, unpolished flint; those of the latter were more skilfully made, and were sometimes very highly finished.

Relics of all these periods have been found in Devonshire, but it is not always possible to say to which age a particular weapon or implement belongs. The use of flint for arrow-heads, for example, certainly continued long after the invention of bronze; and bronze was employed, especially for ornaments, long after iron had come into regular use.

Traces of Palaeolithic man are nowhere common, and they are rare in Devonshire; but they have been found, in the shape of massive tools of roughly chipped flint, in Kent's Cavern near Torquay, in the Cattedown Cave, now destroyed, near Plymouth, and in the river-drift near Axminster. Very fine examples have been obtained from a ballast-pit at Broome.

Remains belonging to the Neolithic Age, on the other hand, are not only much more abundant, but are of greater variety, showing an advance in knowledge of the arts of life, and they include not only axe-heads,

spear-points, and arrow-tips of flint, in some cases beautifully finished, but weapons and ornaments of bone and
horn, rude pottery, spindle-whorls, and even primitive
musical instruments. No traces of the dwellings of the
people of this age, except as regards the caves that
sheltered some of them and their predecessors of the earlier

Dolmen near Drewsteignton

time, have yet been found, nor of their graves, if we except
the Spinster's Rock, two miles west of Drewsteignton.
This is a dolmen consisting of three huge stones, on
which rests a still larger block, twelve feet long and
estimated at sixteen tons in weight; it doubtless was a
Neolithic burying-place.

Relics of the Bronze Age are far more abundant,

and they form, indeed, the chief archaeological feature of Devonshire. Dartmoor in particular, chiefly perhaps because its great upland wilderness has never been broken up by the plough, is dotted all over with remains of primitive Bronze Age dwellings, sometimes standing alone, sometimes grouped in villages and surrounded by a wall, and also with many stone circles, tumuli, kistvaens, menhirs, and rows of upright stones. It is, moreover, intersected by a network of ancient trackways, linking settlement to settlement.

**Palstave of the Bronze Age**
(*In Exeter Museum*)

The best example of an early Bronze Age village is Grimspound, eight miles north-east of Princetown. It consists of twenty-four round huts made of stone slabs set on end and standing about three feet above ground, scattered over a space of about four acres, surrounded by a nearly circular double wall, from nine to fourteen feet thick, and about five feet high, built of blocks of granite, some of which are tons in weight. Half the huts contain fire-hearths, which have been much used; and a good many have raised stone benches, from eight inches to a foot high. Except that the roofs, which were no doubt made of poles and thatch, have disappeared, these primitive dwellings, of which there are hundreds on Dartmoor, are probably much in the same condition as when they were inhabited, perhaps 2000 years ago. It has been suggested that Grimspound, like many ancient camps or hill-forts, was a place of refuge in times of danger, rather than a permanently occupied fortress.

Fernworthy Circle, near Chagford

The objects found in these and similar dwellings consist of flint implements, pottery—some of it decorated—spindle-whorls, and cooking-stones such as are still in use among the Eskimo. So far, no fragment of metal of any kind has been discovered, from which it might, perhaps, have been inferred that the huts were the work of a race unacquainted with the use of metal. But the pottery so closely resembles that found in the burial-mounds of the Bronze Age that archaeologists are satisfied that the dwellings belong to that period.

Other remarkable monuments are the stone circles,

Hurston Stone Alignment

or upright, unhewn blocks of granite arranged in rings, of which there are many on Dartmoor, and of which some of the finest are the Grey Wethers near Post Bridge, and similar structures near Chagford and on Langstone Moor; the avenues or alignments, consisting of rows of stones set up in straight lines, as at Merivale Bridge, Hurston, and Challacombe ; and the menhirs, which are single stones, sometimes twelve feet high, of which the best are at Drizlecombe, at Merivale, and on Langstone Moor.

Triple Stone Row and Circle near Headlands, Dartmoor

It is probable that these structures were connected with primitive forms of worship, but Sir Norman Lockyer has endeavoured to show, in one of the most fascinating chapters of archaeological research, that the circles and avenues, and the monoliths or menhirs connected with them, were in all probability set up as rough astronomical instruments for observing the rising of the sun or of particular stars, in order to regulate the true length of the year. It is even possible, he contends, to form some idea of the date of their erection. Thus the two avenues at Merivale were probably laid out at different times, one about 1610 B.C., and the other about 1420 B.C., in order to watch the sunrise in May, which was then regarded as the first month of the year; while the avenues at Challacombe, among the most remarkable of all the monuments, are probably of far older date, perhaps 3500 B.C., and seem to have been arranged for the observation of sunrise in November, a month long accepted by some Celtic tribes as marking the beginning of the year.

Kistvaens, of which nearly 100 have been found, almost all of them on Dartmoor, are small stone burial chambers, generally used for the reception of the burnt ashes of the dead, probably in many instances originally covered with earth, and made of four slabs of granite set on edge, forming a sort of vault, with another and more massive stone laid on the top. Specially good examples have been found at Fernworthy, on Lakehead Hill, and at Plymouth. The most remarkable kistvaen —belonging, however, to a later period, when burial had displaced cremation—was that discovered on Lundy, containing a human skeleton eight feet two inches in length.

Tumuli or burial mounds, called cairns when they are made of small stones, and barrows when they are merely piles of earth, are to be seen in all parts of Devonshire, especially on high ground, sometimes alone, sometimes in groups of from two to ten or twelve, and always round. No long barrows have been discovered in the county.

The men of the Bronze Age reached a much higher stage of civilisation than their stone-using predecessors, and very fine examples of their bronze swords, daggers, spear-heads and axes; of their pottery, some of it finely decorated; and of their ornaments, including beads of shale and clay and amber, and an amber dagger-hilt with studs of gold, have been found at various places in Devonshire.

Relics that can be attributed with certainty to the prehistoric Iron Age are rare, partly, no doubt, because iron so quickly rusts away. Some very remarkable remains of this period were however found on Stamford Hill near Plymouth, during the construction of a fort, when the workmen dug into an ancient burial-ground, in which, in addition to human bones, were discovered red, black, and yellow pottery, mirrors and finger rings of bronze, fragments of beautiful amber-tinted glass, and some much-corroded cutting-instruments of iron. In the old camp called Holne Chase Castle, a man digging out a rabbit came upon about a dozen bars of rusty iron, two feet long, nearly two inches broad, and a quarter of an inch thick, which although at first mistaken, as has been the case elsewhere, for unfinished sword-blades, were subsequently identified as specimens of the iron currency-bars which passed as money among the ancient Britons. British coins of gold, silver, and copper have been found in several places, particularly at Exeter and at Mount Batten near Plymouth. The commonest are of what is known as the Channel Islands type, bearing the effigy of a horse rudely imitated from the Macedonian stater, and probably struck about 200 B.C.

A special feature of the prehistoric antiquities of Devonshire is the great number of camps or hill-forts, of which there are more than 140 ; some in the heart of the county, some on the coast, often on prominent headlands, and some along the lines of division between this and the adjoining shires ; showing in many instances great military skill and knowledge both in construction and in the choice of good, defensive sites.  Two of the most remarkable of the many strongholds, both near Honiton, are the great encampment of Dumpdon and the magnificently planned fortress of Hembury—a monument of military skill.  Another fine example is Hawkesdown, the strongest of the chain of border forts— of which Membury and Musbury are two important links—built along the river Axe as defences against the ancient inhabitants of Dorsetshire.  One of the largest and most elaborate of all is Clovelly Dykes, twenty acres in extent, and defended by from three to five lines of intricate earthworks.  Another remarkable fort, and the largest in the county, is Milber Down, two miles south-east of Newton Abbot.

There is no real clue to the makers of these fortresses.  Roman coins have been found in several of them, but there is not one which competent authorities attribute to the Romans, and it is probable that they were built by British tribes during the ages of Bronze and of Iron, but perhaps chiefly by the Gaels or Goidels.

A few of them have played a part in modern history. Cadbury, for instance, was occupied by Fairfax in 1645;

**Bronze Centaur forming the Head of a Roman Standard**

(*Found at Sidmouth*)

and in 1688 the Prince of Orange parked his artillery in the great fort of Milber Down, a camp which it is thought was not only occupied but adapted by the Romans.

The Romans, as was pointed out in the previous chapter, left comparatively few traces in Devonshire; and at only twelve spots in the county have Roman relics other than coins been discovered. At Exeter, the only town where there seems to have been continuous occupation, there have been found the foundations of the city walls, a bath, several tesselated pavements—one of which has been relaid in the hall of the police-court, on the spot where it was discovered—statuettes in bronze and in stone, engraved gems, lamps—one encrusted with lizards, toads, and newts—many pieces of finely decorated Samian pottery, and great numbers of coins. The finest mosaic pavement was that found at Uplyme, on the site of one of the only two known Roman villas in the county. Coins have been dug up in many places. Considerable hoards have been discovered at Compton Gifford, Holcombe, Tiverton, and Widworthy. The largest, however, was at Kingskerswell, where 2000 were found together. Perhaps the most remarkable Roman relic was a bronze object found on the beach at Sidmouth, believed to be the head of a standard, and representing Chiron the Centaur carrying Achilles.

Except as regards coins, of which great numbers are in existence, especially, as has been stated, in the Royal Museum at Stockholm, few antiquities that can be ascribed to the Saxons have been found in Devonshire. One remarkable relic, now in the British Museum, was a bronze sword-hilt, dug up in Exeter in 1833, finely ornamented with key-pattern, and inscribed LEOFRIC · ME · FEC(IT). Part of the lettering is inverted, a fact which at first entirely disguised the real words. One of the treasures of Exeter Cathedral is an Old English manuscript headed "A Mycel Englisc Bok," placed in the Chapter Library by Leofric, who in 1050 was bishop of the diocese. It contains some of the

Saxon Sword-hilt

work of Cynewulf and his school. One poem has been thus translated :

"To the Frisian wife
comes a dear welcome guest;
the keel is at rest;
his vessel is come;
her husband is home;
her own cherished lord
she leads to the board;
his wet weeds she wrings;
dry garments she brings.
Ah, happy is he
whom, home from the sea
his true love awaits."

A Norman relic of the highest interest is the Exeter Domesday Book, also preserved in the Chapter Library, describing in greater detail than is given by the Winchester Survey, especially as regards live stock, the five counties of Devon, Cornwall, Dorset, Somerset, and Wiltshire. Among many points of difference between the two books is that where the more general survey gives the letters T. R. E., for *tempore regis Edwardi*, meaning that such were the facts in the time of Edward the Confessor, the Exeter Book uses the phrase "eadie qᵃ rex · E · f · u · & · m ·" for *eâ die qua Rex Edwardus fuit vivus et mortuus*, that is to say, "on the day when King Edward was alive and dead," meaning the day he died, which was the 6th of January, 1066. It is worthy of note that, with one exception, the Devonshire Hundreds were the same in 1086, the supposed year of the completion of the Survey, as they are in our time.

Scattered up and down over Devonshire are many old stone crosses, some in churchyards or by the wayside, probably intended as preaching places, and some standing on the open moor as marks of boundary or of lines of ancient roadway, a few of them bearing brief inscriptions or traces of decoration. Fine examples are those of Addiscot, Helliton, Mary Tavy, and South Zeal, the last-named of which has been restored, and measures, with its steps, eighteen feet in height; the Merchant's Cross near Meavy, the tallest on Dartmoor; the very ancient Coplestone Cross, once decorated with interlaced Celtic ornament, standing where three parishes meet, and named in an Anglo-Saxon charter of 974; and the Nun's or Siward's Cross, inscribed on one side (SI)WARD, and on the other BOC LOND, set up in the twelfth century to mark the boundary between the royal forest and the property of Buckland Abbey.

With these may be mentioned the inscribed stones of Lustleigh, Stowford, Tiverton, and Fardel—the latter now removed. The last-named, and one of the three Tiverton examples, bear inscriptions in Ogham or Irish runic characters.

Cyclopean Bridge, Dartmoor

Among the features of Dartmoor are the so-called "clapper bridges," made of great blocks of unhewn stone. Their date and their builders are matters of conjecture. If, as has been said, they were meant for packhorse traffic, they may be at least as old as the time of the Norman Conquest, for packhorses are mentioned in the Exeter Domesday. There are fine examples at Dartmeet, Bellaford, and near Scaurhill Circle. But the most striking and perhaps the most ancient is that at Post Bridge, in which two of the stones are fifteen feet long.

The idea has long been abandoned that Logan or rocking stones were the work of man, or that the rock basins which are found on the top of some of the Dartmoor tors had anything to do with Druidical ceremonies. Both are now recognised as of natural origin. But it may here be noted that the largest of the Devonshire Logans is the Rugglestone, at Widecombe, estimated at from 100 to 150 tons, but it can no longer be moved. The largest that will still rock is one of about fifty tons weight at Smallacombe. The largest of the many rock basins, which vary from a few inches to five feet or more, are on Heltor and Kestor.

## 20. Architecture—(*a*) Ecclesiastical.

The ecclesiastical buildings of Devonshire,—its magnificent cathedral, which is without doubt the finest example of the Decorated style in all England ; its many noble churches, some of which, specially remarkable for their interest and beauty, are situated in remote and thinly-peopled rural parishes ; and, in a minor degree, the picturesque fragments of its ruined abbeys—form altogether one of the most striking features of the county. Speaking generally, it may be said that Devonshire churches, as a whole, are remarkable for their interiors, very many of them containing beautiful wood-work, especially rood-screens, and finely-carved stone pulpits, to which, in many instances, the addition of gold and colour has lent a still more striking and even gorgeous effect. Some of the exteriors also are very beautiful ; but, on the other hand, many of them, partly on account of the intractable nature of the stone of which they were built, are simple and even severe in character. As might be expected, the material varies with the geological formation. Thus, many churches were built of grey limestone. In East Devon much use was made of flints and of freestone from Beer. Round Exeter and Crediton volcanic tufa was often employed, particularly in vaulting. The use of Old Red Sandstone and even of granite greatly affected the style, which in buildings of those difficult materials is plain and with little ornament.

A large proportion of the churches of Devonshire were more or less rebuilt during the Perpendicular period, that is, between 1377 and 1547; but many, probably even the majority of them, contain features of earlier dates, in a few cases going as far back as Saxon times. Some have been skilfully restored. But in too many cases the work of renewal was carried out in an age when church architecture was imperfectly understood, and when the value of old and beautiful, even if time-worn, details was not sufficiently appreciated ; and it is unfortunately true that, in order to accomplish needless or barbarous alterations, many interesting features were ruthlessly swept away.

The oldest existing work is the Saxon masonry in the bases of some of the central Norman towers—those of Branscombe, Axminster, and Colyton, for example, and in the crypt of Sidbury. No church is wholly or even

Norman Doorway, Axminster Church

largely Norman ; but the transeptal towers of Exeter
Cathedral are of this period, as are the towers of South
Brent, Ilfracombe, and Aveton Gifford, in addition to
those named above.  There are also fine Norman door-
ways at Paignton, Kelly, Axminster, Hartland, Bishop's
Teignton, and elsewhere.  In at least a hundred churches,
most of which probably possess no other feature of the

Ottery St Mary Church

time, there are Norman fonts, of which the most remarkable are those at Hartland, Alphington, and Bradsworthy.
The font in Dolton church is believed to be Saxon.

Perhaps the best examples of Early English architecture are to be seen in the aisles and transeptal towers—
the latter imitated from those of Exeter Cathedral—of the
very beautiful church of Ottery St Mary, the finest and

most interesting church in Devon.   The plain little building on Brent Tor, one of the smallest of churches, measuring only forty feet by fourteen, is probably all Early English.   The churches of Sampford Peverell, Haccombe, and Aveton Gifford are almost entirely of this period, as are the transepts and central tower of Combe Martin and

Decorated Window, Exeter Cathedral

the tower of Buckfastleigh, which carries one of the few spires in the county.

The Decorated style is not so well represented as regards the number of examples.   But to this period belongs almost the whole of Exeter Cathedral, a great part of the beautiful church of Tavistock, and the nave, chancel, and Lady-chapel of Ottery St Mary.   There is also good work of this style at Beer Ferris, Plympton, and Denbury.

So many Devonshire churches, as already remarked, were rebuilt in Tudor times that the majority of the ecclesiastical buildings in the county appear to belong to the Perpendicular period. There is very beautiful Perpendicular work in the church of Tiverton, whose south front and chapel were decorated by their founder, a wool-merchant named Greenway, with very elaborate carvings, some symbolic of his trade, and some representing scenes from the life of Christ. Other notable churches mainly of this period—to name a few only out of a multitude of examples—are those of Crediton, Hartland, Plymptree, Awliscombe, Kenton, Harberton, Dartmouth, and Buckland Monachorum.

To the Perpendicular period belong the finest of the Devonshire towers, which as a rule, however, owing in many instances to the absence or to the poorness of buttresses and pinnacles, lack the majesty of those which are so striking a feature of the ecclesiastical architecture of Somerset. There is a group of three towers in near neighbourhood, assigned by tradition to the same architect, and known as Length, Strength and Beauty, at Bishop's Nympton, South Molton, and Chittlehampton, respectively ; and the last of these, a magnificent piece of architecture, is the most beautiful specimen of an enriched tower in Devonshire. Other very fine towers are those of Cullompton, Chumleigh, Berrynarbor, Arlington, Kentisbury, and Combe Martin. The tower of Colyton is unique in character, being crowned by an octagonal lantern supported by slender flying buttresses.  There are not now many spires in Devonshire, but there are fine examples at Modbury—which tapers the whole way up— and at Barnstaple, both of the sixteenth century ; and there are others at Braunton, Brushford, and West Worlington.  One of the towers of Ottery St Mary carries a spire, the other is without.

One of the special characteristics of Devonshire churches is their woodwork, their roofs and bench-ends, their pulpits—although some of the best of these are of stone—and, above all, their rood-screens.  The last-named are among the finest in the kingdom, and are not rivalled even in Norfolk and Suffolk.

There are good timber roofs at Cullompton, Widecombe, South Tawton, Hartland, Ashburton, Chittlehampton, Sampford Courtenay, and Hatherleigh.  The bench-ends at Abbotsham, Ilsington, Ashton, Mortehoe, Tawstock, Braunton, Monksleigh, Frithelstock, East Budleigh, and Combe-in-Teignhead are specially fine. The seventeenth century seats at Cruwys Morchard are inscribed with the farm names of the parish.

Rood-screens, which are here the most remarkable feature of the Perpendicular period, are very numerous in Devonshire.  Although many have disappeared, having been removed or broken up, there are still some 150 in more or less perfect condition.  So many of them, moreover, are of such truly exquisite workmanship that it is difficult to say which are the most beautiful.  The material, in the majority of cases, is wood, perhaps because of the scarcity of tractable stone—elaborately carved, and

very often splendidly decorated with gold and colour.
There are, however, magnificent screens of stone in
Exeter Cathedral and in the churches of Totnes and
Awliscombe.

It is probable that most of the screens were the work
of native craftsmen, but there are some whose style shows
distinct signs of foreign influence. The beautiful screen at

Rood Screen and Pulpit, Harberton Church

Harberton, for example, suggests Spanish work or influence,
that of Colebrook French, that of Kenton Flemish, and
that of Swymbridge Italian. While by far the greater
number are of the Perpendicular period, that of Washfield
is Jacobean, and that of Cruwys Morchard is Georgian.
It is perhaps generally considered that the magnificent
screen at Kenton is the finest of all ; but it has a good

many rivals which closely approach it in beauty of design and in excellence of workmanship. Other splendid specimens, all of them of great beauty, are those at Kentisbere, Hartland, Hemyock, Swymbridge, Kingsnympton, Dartmouth, Honiton, Holbeton, Tawstock, Lustleigh, Lapford, Pinhoe, and Uffculme. The last-named, which measures sixty-seven feet, is the longest, and that at Welcombe is believed to be the oldest, in the county.

Carved pulpits are another special feature of Devonshire churches. The finest stone pulpit, which is at Harberton, contains, like the beautiful examples at South Molton and Chittlehampton, full-length figures in panels. Other good stone pulpits are at Pilton, Totnes, Paignton, Dartmouth and elsewhere. Two particularly fine carved oaken pulpits are those of Hartland and Kenton, the latter of which is very richly decorated with gold and colour. There are also good specimens at East Allington, Tor Bryan, Ipplepen, and Holne.

Ancient stained glass is very rare in Devon, much having been destroyed by Puritan fanatics. The best which has survived is at Doddiscombeleigh, where there are four very beautiful windows. There is also very good glass at Cheriton Bishop and Budleigh ; and some less striking but noteworthy examples may be seen at Ashton, Christow, Cadbury, Manaton, Atherington and other places. The great east window in Exeter Cathedral contains some very fine coloured glass, and there are a few remains in some of the clerestory windows.

Among the very striking recumbent effigies of warriors

and churchmen and great ladies to be found in our churches there are some not to be surpassed in England; and the magnificent examples in Exeter Cathedral, in particular, afford most valuable chronological studies both of costume and of carving. Among the finest of those in the cathedral are the splendid thirteenth century alabaster effigy of Bishop Bronescombe, the fourteenth century mail-clad figures of Humphrey de Bohun and Sir Richard Stapledon, and the sixteenth century effigy of Bishop Hugh Oldham. Perhaps the grandest of those in parish churches, to name a few only out of many, are the fourteenth century effigies of Sir Otho Grandisson and his wife at Ottery St Mary, and the seventeenth century figures of Denys Rolle and his wife at Bicton. Other fine effigies are at Paignton, Broadclyst, Landkey, Tawstock, Haccombe, and Horwood and the Seymour tomb at Berry Pomeroy.

Our county is rich also in monumental brasses, most of which date from the fifteenth and sixteenth centuries. Among the finest are the splendid triple-canopied fifteenth century brasses in the church of St Saviour, Dartmouth, in memory of Sir John Hawley and his two wives, the fifteenth century brass of Sir Nicholas Carew in Exeter Cathedral, which is a wonderful representation of the military costume of the period, and the fine sixteenth century brass at Tiverton, in memory of John and Joan Greenway. Other good brasses are in Exeter Cathedral and in the churches of Stoke-in-Teignhead, Stoke Fleming, Clovelly, Braunton, Haccombe, and Clyst St George.

Many of the churches, that of Pilton, for example, have specially musical peals of bells; and there are very

The Seymour Tomb, Berry Pomeroy Church

old bells at Ogwell, Abbot's Beckington, Alverdiscott and Hittesleigh. At the latter place is the most ancient bell in the county.

By far the most important and remarkable ecclesiastical building in Devonshire is Exeter Cathedral, which, although not one of the largest or—externally, at any rate—one of the most majestic cathedrals in this country, is without doubt the most beautiful example in all England of the Decorated style of architecture. It is built of Beer stone; a material which when first quarried is white and easily worked, but which, in this case, is now dark and crumbling with age.

The two features which specially distinguish this from all other cathedrals are its transeptal towers, of which the only other English example is at Ottery St Mary; and the great length of its roof, which extends unbroken over nave and choir.

The exterior is further remarkable for the statuary on its west front with its figures of kings and knights, saints and angels; for its flying buttresses and its richly carved pinnacles. And the interior, which has been called the finest in Europe, is distinguished for the beautiful tracery of its many Decorated windows, the elaborate details of its side-chapels; its episcopal throne, more than fifty feet high, a marvel of wood-carving, without a rival in the island; its noble screen, one of the best in a county particularly rich in screens; its ancient and quaintly-carved misereres, the earliest in England; its fourteenth century minstrels' gallery, the most nearly perfect known; its long stretch of stone vaulted roof, the longest of the kind in existence; its clustered columns; its richly yet delicately

Exeter Cathedral, West Front

carved bosses and finely sculptured corbels; its many
monuments and recumbent effigies of knights in armour
and of bishops in their robes of office; and, generally, by
the wonderful uniformity and symmetry of its design.

The cathedral is the work of many hands. Hardly
one of its long line of bishops but has left his mark upon
it. But that it is an architectural masterpiece is due in
the first place to the genius of one man, and in the second
place to the wisdom of his successors in faithfully carrying
out his original design.

The cathedral stands on the site of a Saxon church, of
which no trace remains; and of the Norman edifice
which succeeded it little is left but the two transeptal
towers. These towers, the northern of which has been
much altered, and now has strongly-marked Perpendicular
characters, were built by Bishop Warelwast (1107–1136),
the son of William the Conqueror's sister; and Bishop
Marshall (1194–1206), brother of that Earl of Pembroke
who helped Henry II in the conquest of Ireland, finished
the building in the Norman style. But it was Bishop
Quivil (1280–1291) who planned the reconstruction of
the whole cathedral in the Decorated style, with the
exception of the towers, and himself began the trans-
formation, rebuilding the transepts, the Lady-chapel, and
part of the nave. And although the work extended over
more than a hundred years after Quivil's time, and was
continued far into the Perpendicular period, the architecture was not altered, and there are few features in the
building which are not in keeping with the bishop's first
design. The magnificent Perpendicular eastern window

is filled with beautiful glass of the previous period, and it is believed that its tracery also was originally of the Decorated style.

Of the bishops who succeeded Quivil, Stapledon (1308–1326), a statesman as well as a prelate and an architect, murdered in Cheapside by ruffian partisans of the She-Wolf of France, carried out some of the finest

The Nave, Exeter Cathedral

work in the building, including the rood-screen, the episcopal throne, and the stone sedilia.   Last of the great builders was Bishop Grandisson (1327–1369), who, in his long tenure of the see, completed and finally consecrated the cathedral, to which, however, some details were added by those who followed him.   Bishop Brantyngham (1370–1394), for instance, finished the west front, the great east window, and the cloisters—destroyed by the Puritans and only recently rebuilt. There have been two main restorations of the cathedral; one in 1662, and one between 1870 and 1877, when the reredos and other features were added.

The chapter-house, which is an exception in style to the rest of the building, its lower part being Early English, and its upper part Perpendicular, contains part of the cathedral library. The rest of the 15,000 volumes of books, together with some very valuable manuscripts, including the Exeter Domesday Book, Leofric's Book of Saxon Poetry, and the original charter signed by Edward the Confessor, Earl Godwin, Harold and Tostig, authorising the removal of the see from Crediton to Exeter are preserved in the Chapter Library.

In the north transept of the cathedral are the dials of an ancient and curious clock, believed to have been set up early in the reign of Edward III, although its movement has been renewed. It sounds the hours and the curfew on Great Peter, a ponderous bell in the tower above it. The peal of ten bells in the southern tower is the heaviest in England.

At many points in Devonshire may be seen the ruins of monasteries, priories, and nunneries which were closed by order of Henry VIII. Almost all of them have suffered so severely from decay, or perhaps even more from having been used as quarries, that in a great many instances only a few fragments of ruin remain of what, in their time, were large and magnificent buildings.

These monastic houses were originally founded as

places to which people might retreat who wished to retire from the world, and to lead simple lives of holiness, benevolence, and poverty, serving God and benefiting their fellows. For a time the inmates did all these things. As long as they were poor they were a blessing to the countries where they lived. They preached to the people, they taught in schools, tended the poor and the sick, practised agriculture and many useful arts, such as the construction of clocks, keeping alive such learning as there was, and making beautiful manuscript copies of the Bible and of the works of classical writers which otherwise would have been lost. But when they grew rich they became idle, careless, and ignorant, and their lives too often a scandal to the world. Henry VIII, as the result of a commission which he sent round to enquire into their condition, decided to suppress them. The houses were closed, their inmates were scattered, their estates were sold for trifling sums or given to the king's favourites, while part of their vast wealth was used in founding grammar-schools.

The richest monastic house in the county was the Cistercian Priory of Plympton, of which little now remains beyond the refectory and the kitchen. Of the Benedictine monastery of Tavistock, an establishment second in wealth only to Plympton, the gateway, a porch, and two towers alone are left. The remains of the Norbertine abbey of Torre consist chiefly of the refectory, a gate-house, and the fine building known as the Spanish Barn, from a tradition that Spanish prisoners of war were confined in it.

The two Cistercian houses of Buckfast and Buckland are specially interesting. The ruins of the former, which was a very ancient and very rich establishment, whose last abbot attained his office as a reward for having helped to capture Tyndale, were bought in 1882 by a community of French monks, who have rebuilt much of the abbey in the original style. Part of Buckland Abbey, which had

Buckland Abbey

been converted into a dwelling-house in Henry VIII's time, was bought and rebuilt by Sir Francis Drake. Several relics of Drake are preserved here, and the house, with its fine cedars and stately tulip trees, is one of the most picturesque buildings in Devon. Hartland Abbey, originally founded, like Buckfast, in Saxon times, has also been converted into a dwelling-house, into which were built the Early English cloisters. At Leigh, near

Christow, are the very picturesque remains—a fine gate-house, the refectory, and the dormitory—of a small cell connected with Buckland Abbey.

Other monastic remains, mostly in a fragmentary condition, are those at Polsloe (Benedictine nuns), Denbury (Benedictine cell connected with Tavistock), Newenham and Dunkeswell (both Cistercian), Cornworthy (Augustinian nuns), and Frithelstock (a house of Augustinian canons).

## 21.  Architecture—(*b*) **Military.**

As has already been pointed out, there were in Devonshire a very great number of primitive castles or fortresses, generally on the tops of hills, and consisting simply of enclosures surrounded by ramparts of earth or of loose stones. After the Norman Conquest castles of a very different type, strongly built of stone, were erected in our county, as in many other parts of England, partly by order of the king himself, and partly by his knights and nobles, who found it necessary to defend themselves against the Saxons, of whose lands they had taken possession. By the end of the reign of King Stephen, after less than ninety years of Norman rule, there were 1115 such strongholds in England.

The early Norman castle consisted, as a rule, of a single three-storied tower with walls of great thickness. But at a later period, after the experience gained in the Crusades, and in consequence of the introduction of powerful engines capable of throwing great stones against a besieged fortress, military architecture became much more elaborate.

The castles of the Middle Ages, which sometimes occupied a space of many acres, were usually built on high ground, or close to a river or the sea-shore, and were almost always surrounded by a ditch or moat, which, if possible, was filled with water. Inside the moat was a high and very thick wall, generally with towers at intervals, especially at the corners, with a parapet to shelter the men fighting on the top of it, and with spaces called embrasures through which they could shoot arrows at the enemy. The principal gate was strongly defended by covering towers, and above it were holes through which melted lead, or boiling water, or hot pitch or sand could be thrown on the besiegers. The doorway was reached by a drawbridge, raised and lowered by chains, and was closed by a heavy door, or a strong grating called a portcullis. Smaller gates were the postern and the sally-port.

The space inside the outer wall was known as the outer bailey, inside which was another wall, also with towers and a gate, within which were dwellings and store-houses. This was the inner bailey; and within it was the most important part of the fortress, the high tower called the keep—a building of several floors, with walls 15 or even 20 feet thick—the last place of retreat when the rest of the castle was taken. On the ground floor, which had no windows, were the well, sometimes of immense depth, the dungeon, and the store-rooms.

On the next floor, which was lighted by narrow loop-holes, were the soldiers' quarters.  On upper floors were the chapel and the apartments of the governor and his family.

There were nearly twenty Norman and mediaeval castles in Devonshire, few of them large or elaborate from the military point of view, and some, perhaps even the majority of them, insignificant in size and simple in style.  Of some of these twenty strongholds no trace remains.  Almost all, of which anything survives, are ruinous and uninhabitable.  And although a few have been partially restored and are now occupied as dwelling-houses, only one, that of Powderham, retains its ancient dignity, after continuous occupation that has lasted for nearly six centuries.  Comparatively little is known of the history of these castles ; but many of them have been the scenes of fighting, especially in Norman times, in the reign of Henry VIII and during the Civil War.

One of the most famous and at the same time one of the oldest of Devonshire castles, is that of Exeter, called Rougemont from the colour of the rock on which it stands, built in 1067 by William the Conqueror on the site of an earlier fortress constructed by Athelstan, destroyed by Sweyn, and restored by Cnut and Edward the Confessor.  Its first Norman governor was Baldwin de Brioniis, the Conqueror's nephew by marriage.  It, or the city round it, has sustained many sieges ; by the Saxons, by the sons of Harold, by King Stephen, by the Yorkists, by rebel armies, by Royalists and Parliamentarians, but it was ruinous even when Fairfax captured

Exeter in 1646. A mere fragment, consisting chiefly of the gateway tower, is all that now remains of it.

The picturesque ruins of Okehampton Castle, built, it is believed, in the thirteenth century, partly of water-worn stones from the river below, and dismantled by order of Henry VIII, stand in a strong position above the West Okement, and include the remains of many rooms and of a great banqueting hall. A French prisoner of war has left a Latin inscription cut in one of the stones.

The remains of the ancient castle of Berry Pomeroy—founded, as some think, in the reign of William I—the most picturesque ruin in Devonshire, standing on a rocky eminence surrounded by dense woods three miles north-east of Totnes, consist principally of two towers, and of the gate-house and the chapel, whose fine masonry, much of which is thirteenth century work, is overgrown with moss and ivy. Inside the original building stands a very large but unfinished mansion, of great magnificence, begun by Lord Seymour, to whom, when Sir Thomas Pomeroy was deprived of his estates because of his share in the Commotion of 1549, the castle was given. The property still belongs to the Duke of Somerset, and has thus been in the hands of only two families since the Norman Conquest.

Compton Castle, about two miles west of Torquay, a very strongly fortified manor-house of the early fifteenth century built on the site of a castle of William I's time, is specially interesting from its association with Sir Humphrey Gilbert. It is now occupied as a farm. Nearness to the

Compton Castle

sea, and the consequent danger from the raids of foreign invaders, led to the strength of its defences. The building, which is large and picturesque, includes a number of ancient features, especially the chapel and two gateways.

Lydford Castle, now little more than the shell of a square tower, stands on an artificial mound, and is said to have been founded soon after the Conquest. It was a place of great importance in the palmy days of Devonshire tin-mining, when Lydford was one of the chief towns in the county. Here, from a very early period until late in the eighteenth century, was held the Stannary Court, proverbial for its arbitrary methods of procedure ; and within the walls was a notorious dungeon, used as the Stannary prison.

Plympton Castle, built by Richard de Redvers, first Earl of Devon, and the scene of fighting in the reigns of Stephen, John, and Charles I, is now quite ruinous. Tiverton Castle, ascribed to the same founder, but reconstructed in the fourteenth century and dismantled after the Civil War, has been partly adapted as a modern house. It sustained a brief siege, of only a few hours' duration, by Fairfax in 1645. A cannon-shot cut the chain of the drawbridge, the bridge fell, and the besiegers, pouring in, were quickly masters of the fortress. The chief ancient features are the great gate, a tower, and the remains of the banqueting hall and the chapel. The scanty remains of Hemyock Castle, two miles east of Culmstock, and not far from the border of Somerset, at first garrisoned for the Parliament, then taken by the Royalists, and finally dismantled by Cromwell, consist of little more than the gateway and its covering-towers, which are of flint. Totnes Castle, whose ivy-clad walls of red sandstone look down upon the river Dart, was founded by Judhael, soon after the Conquest, but it has been a ruin since the time of Henry VIII. Of Dartmouth Castle, a very picturesque ruin at the end of a promontory guarding the harbour, the chief remains are a square tower of the time of Edward IV, and a round tower of the reign of Henry VII. The place may still be seen where a chain was drawn across the river to Gomerock Castle, a small fort on the opposite shore, to keep hostile ships from sailing up the Dart. Kingswear Castle, a small thirteenth century building on the same river, the scene of some fighting during the Civil War, has been restored, and is now a private residence. Salcombe or Clifton Castle, on the Kingsbridge estuary, one of Henry VIII's coast defences against the long-expected attack of the Spanish Armada, was the last place on the Devonshire mainland to hold out for Charles I.

The square Morisco fortress on Lundy, whose plain walls now shelter cottages that have been built inside it, is twelfth century work. The scanty ruins of Colcombe Castle near Colyton, supposed to have been destroyed by the Parliamentarians, and the square tower of Gidleigh, not far from Chagford, date, it is believed, from the century following; and the castle of Ilton, two miles north-west of Salcombe, on the Kingsbridge estuary, now used as a farm, was built in the fourteenth century. Of Torrington Castle a few fragments only are left. The castles of Exmouth and Bampton have entirely disappeared ; and of Barnstaple Castle, built, it is said, by Athelstan, but ruinous as far back as the reign of Henry VIII, nothing but the site remains.

Powderham Castle, the only one of all these fortresses which has been continuously inhabited since its foundation, stands—from a military point of view—on a poor site, on low ground close to the estuary of the Exe. Its chief charm is in its setting, in its beautiful park and fine timber, especially its magnificent oak-trees. Built in 1325 in the form of a long parallelogram, with six towers, four of which remain intact, while two have been restored, it has been altered and added to by many hands, and is now a vast, irregular pile of buildings. Its present owner is the lineal descendant of its original founder, the Sir Philip Courtenay who, in 1367, was knighted by the Black Prince on the field of Navarete. Successfully held for the King in December, 1645, against Fairfax himself, it was taken by Colonel Hammond in the following January after some sharp fighting, in which the Parliamentary troops, as happened on not a few occasions during the war, seized and fortified the village church.

## 22. Architecture—(c) Domestic.

Scattered up and down over Devonshire are many fine old manor-houses, some of them, in parts at least, very ancient, some with picturesque and striking features, many set in very beautiful surroundings, and others of interest for the sake of their historic associations.  Such houses are so numerous that only a few of them can here be even lightly touched upon.

Not one of the famous houses of Devonshire is entirely, or even in great part, as old as the thirteenth century, although there are several that contain features of that

An Old Devon Farmhouse Chimney Corner

period.  Such, for example, is Bowringsleigh, near Kingsbridge, a fine old building, which although mainly Tudor, and containing details of later eras—beautiful Jacobean oak screens and highly-decorated plaster ceilings of the time of William and Mary—has some striking thirteenth century work in it.

K. D.                                    13

At Little Hempstone, near Totnes, is a very interesting and well-preserved pre-Reformation parsonage or priest's residence of the fourteenth century; and Ayshford Court, near Burlescombe, a fine old house now used like the Little Hempstone parsonage as a farm, contains a fourteenth century chapel. Of the same period are the great hall, now dismantled, and the old kitchen and other buildings connected with the mansion of Dartington, which although as a whole a noble example of Elizabethan architecture, was originally erected in the reign of Richard II.

Manor-houses of the fifteenth century are much more numerous. The most remarkable of them—indeed, the finest of all the many great houses in the county, is Wear Gifford, on the Torridge, about $2\frac{1}{2}$ miles south of Bideford, a perfect example of an old English manorial residence, built, it is believed, during the reign of Henry VI. Greatly damaged during the Civil War, the house, which stands in a commanding situation with fine timber, especially oak-trees, about it, was for a long period occupied as a farm, and having become much dilapidated, was restored about eighty years ago. It contains many beautiful and interesting details, but the most striking of the original features are the square embattled tower with the fine entrance archway beneath it, and the magnificent hall, rising to the whole height of the building, with richly-decorated oaken panelling and a carved, open, hammer-beam roof which is one of the very finest examples of Perpendicular woodwork in England. Other good specimens of fifteenth century architecture are Wortham, at Lifton, near the border of Cornwall, an almost perfect house of the period; Bradley, near Newton Abbot, a very picturesque building with a fine hall and chapel; and the main fabric of Exeter Guild-hall, which was erected in 1464 though the front is Elizabethan.

The noblest Tudor mansion in Devonshire is Holcombe Rogus, in the village of that name, near Burlescombe, about three miles from the border of Somerset. A good deal of the building is a modern restoration, but many details of the time of its foundation, in the reign either of Henry VIII or of Edward VI, still remain. As in the case of Wear Gifford, the most striking features of the house are the very picturesque tower and gate-house, and the great hall—a magnificent room, more than forty feet long, lighted by two great six-light windows. Some of the rooms are finely wainscoted with curiously carved oaken panelling. Adjoining the building is the original "church-house," consisting of kitchen, refectory, and cellar, where parishioners could cook their food and brew their beer, where the poor received their doles, and where the needs of casual wayfarers were relieved.

Another very interesting Tudor mansion, only part of which, however, is now habitable, and is used as a farm, is Cadhay, at Ottery St Mary. The interior of the house has been a good deal altered, but the exterior is much as it was in the days of Queen Elizabeth. Its most remark-able feature is the inner court, round which the house is built, and in each of whose four sides, over an arched Tudor doorway, is a highly-decorated projecting canopied

13—2

niche.    In these niches are statues of Henry VIII,
Edward VI, Queen Mary, and Queen Elizabeth.

Another very fine sixteenth century house, containing
also some earlier details, is Bradfield, near Uffculme, in
which are a beautiful music-room, a fine banqueting-hall
with good panelling, a minstrels' gallery, and a richly-
carved roof.    Altogether, this is one of the best examples
of domestic architecture in Devonshire.

Hayes Barton : Sir Walter Ralegh's House

Other interesting and noteworthy houses of the period
are Colleton Barton at Chumleigh, Flete House at Holbeton, Hayes Barton at East Budleigh, and Mol's Coffee
House in Exeter.    Flete has been rebuilt, but it is a
fine mansion, whose beauty is much enhanced by its
surroundings and its avenue of cedars.    Hayes Barton,
where Ralegh was born—it was thus he always spelt his

name—and where a table said to have belonged to him is shown, is a rather modern-looking house, plainly built of "cobb"; but its gables, its mullioned windows and its heavy door are characteristic of the time. In Mol's Coffee House, which is one of the sights of Exeter, is an oak-panelled room decorated with the arms of Drake, Ralegh, Monk and others, in which the great Devonshire soldiers and sailors of Armada days were accustomed to meet.

Two particularly interesting seventeenth century mansions are Sydenham House, not far from Tavistock, and Forde House, near Newton Abbot. In the former, which is a specially fine example of the work of the early part of the century, containing also some fourteenth century details, is some very good carved and decorated woodwork, especially in the form of artistic panelling and stately staircases. There are also secret rooms and passages, some of which have been contrived in the thickness of the walls. The house was greatly damaged during the Civil War, when it is said to have been stormed by the troops of the Parliament. At Forde House, which was taken and retaken several times in the struggle between the King and the Commons, the Prince of Orange slept on the first night after his landing at Brixham. Charles I was there twice, in the first year of his reign.

Some good examples of more modern houses are Kingsnympton, in the parish of that name, about four miles from Chumleigh, surrounded by well-wooded grounds, on a commanding eminence looking down on the Taw;

Mol's Coffee House, Exeter

Ugbrooke, near Chudleigh, standing in a deer-park of rare beauty, finely timbered, and most picturesquely varied by wood and hill and water, and where Dryden's grove may still be seen; Rousdon, near Axmouth, built of flint faced with Purbeck stone, and considered one of the most magnificent modern mansions in Devon ; Saltram House, three miles east-north-east of Plymouth, a stately building

**Sydenham House**

in a large and beautiful park ; and Bicton House near Budleigh Salterton, whose trees, brought from all parts of the world, and including a wonderful avenue of araucarias, form one of the finest collections of the kind in Europe.

Other interesting houses are Ashe House near Axminster, the home of the Drake family and the birth-place of the great Duke of Marlborough, partly burnt down during the Civil War and repaired with stones from the ruins of Newenham Abbey, and now a farm-house; Great Fulford, in the parish of Dunsford, about eight miles west of Exeter, owned by a family who have held it since the time of Richard I, stormed by Fairfax in 1645; and the residence in Exeter of the Abbots of Buckfast, a good example of mediaeval architecture.

Some very picturesque old half-timbered houses are to be seen in Exeter, especially in High Street, North Street, and South Street; and there are so many in Dartmouth that the town has been called the Chester of Devonshire. Nor should the fine old almshouses of Tiverton and Exeter be forgotten.

Devonshire possesses a great variety of building stone; and the materials employed have naturally varied, as a rule, according to the geological formation of the district. Some of the best houses are of Beer stone. Some, as has been shown, are of flint. Brick, which when of good colour and quality is an excellent material, has been largely employed. Many cottages, and even whole villages, such as Otterton and East Budleigh, are built of " cobb," which is a mixture of clay and straw.

Thatch, which is still used for roofing, although to a less extent than formerly, has in the past been the cause of many disastrous fires. As recently as 1866 more than 100 houses were burnt down in Ottery St Mary. Nearly the whole of Chudleigh was thus destroyed in 1807. Fires in Crediton, in the eighteenth century, destroyed hundreds of houses. Perhaps the town that has suffered

Dartmouth: Old Houses in the High Street

most severely in this way is Tiverton, where there were very destructive fires in the eighteenth century. One in 1612 consumed almost every house, and in another, in 1598, no fewer than 400 houses were burnt down.

Newton Village

## 23. Communications: Past and Present.

In prehistoric times Devonshire was crossed by a net-work of trackways, some of which are to-day broad and well-kept high roads. Others form those proverbially narrow, awkward, and frequently muddy Devonshire lanes which are so characteristic of the county, having become worn in the lapse of ages so deep below the level of the adjacent country, owing partly to the softness of the ground, and partly to the heavy rainfall, that their high banks, although often very beautiful, completely shut out the view. Others, again, that once served merely to connect one hill-fort with another, have fallen out of use, and are now hardly to be traced.

These roads, probably begun in the Neolithic Period as footpaths, may have been made into tracks for pack-horses in the Bronze Age, and more or less adapted for wheeled traffic by the prehistoric users of iron. Pack-horses, however, usually or frequently in teams of six, were in common use in the county until the middle of the eighteenth century; and although good roads were made across Dartmoor in 1792 there were parts of that wild district where, before the year 1831, wheeled vehicles were unknown. At the present time the total length of all the roads in Devonshire is only exceeded in the county of Yorkshire.

It is generally believed that no Devonshire road was wholly constructed by the Romans, who probably reached the district by the already existing British coast-road from Dorchester. There are some, however, who think that the Fosse Way joined this road and passed through Exeter, going as far as the river Teign. The Romans made no road beyond this point, at any rate; and here, not far from Newton Abbot, they built over the river a bridge of freestone, on whose foundations the modern structure —the third since then—now rests.

Some ancient roads have been abandoned because of their steepness, or because they have been superseded by more convenient ways. Such, for example, are the lane from Crockam Bridge over the Teign to Trusham, and the Lichway (i.e. the way of the corpse) along which, before 1260, the dead were carried for burial into Lydford, crossing the river over Willsworthy Steps, a series of eighteen stepping-stones. One of the most remarkable of these old roads was the great central trackway on Dartmoor, leading from Chagford to Tavistock, 10 feet wide, 2½ feet deep, and made of rough stones with smaller stones above. Although much of it has been destroyed for the sake of its materials, about 18 miles of it still remain. It was during the seventeenth century that the " Moor-stones "—upright monoliths of granite—were set up to serve as guide-posts for wayfarers during the mists that so often cover the moor. One of the most important highways in Devonshire is the great trunk road from London, which enters the county with the Great Western Railway and accompanies it to Exeter, thence making straight for Plymouth, and passing on into Cornwall.

In common with other English counties Devonshire possesses a number of hamlets whose names end in "ford," a syllable which, in words of Saxon origin, means that an old road there passed through the shallows of a stream or river. Such, to give a few familiar instances, are Chagford, Lydford, and Bideford.

Devonshire canals are short and unimportant. The hilly country is not adapted for them ; and such traffic as some of them once enjoyed has been absorbed by the railways. There is, however, a good deal of traffic on the Exeter Canal—constructed in 1566, and therefore one of the oldest ship canals in England and the first lock-canal in the kingdom—but it is worked at a loss. Most of the Bude Canal has been abandoned, and only two miles of it are now in use. The Grand Western Canal, running ten-and-a-half miles eastward from Tiverton, nearly to the Somerset border—all that was ever made of a waterway intended to reach Taunton—the Stover Canal, two miles in length, and the Hackney Canal, only half-a-mile long, both connected with the river Teign, are all under the control of the Great Western Railway. No Devonshire river is now of much value as a waterway. There is some traffic on some of the estuaries, especially the Teign; and the Tamar is navigable to Gunnislake, a distance of twenty miles.

There are in our county some very old lines of stone-tramway for horse-traction; from Tavistock to Princetown, for example, and from the Heytor quarries to the head of the Stover Canal, but they are no longer in use. Down the former was brought granite to build London Bridge.

The railway from London to Bristol was opened by the Great Western Company in 1841, was continued to Exeter by the Bristol and Exeter Company in 1844, and to Plymouth by the South Devon Company in 1846. Atmospheric pressure was tried for a time between Exeter and Newton Abbot, but it was a failure, and was soon superseded by steam-traction.

Brunel's railways were made on the broad-gauge system with seven feet between the rails, in order to give stability to the trains and to allow of a high rate of speed; and the entire line from London to Plymouth was broad gauge. Most other companies, however, adopted the narrow gauge, in which there is only four feet eight-and-a-half inches between the rails; and owing mainly to the inconvenience of not being able to interchange rolling-

Teignmouth : the Coast Line and Sea-wall

stock with other lines, the Great Western Railway Company have converted their whole system to narrow gauge.

The Devonshire railways are now owned by two companies only, the Great Western and the London and South Western. The latter, which enters the county near Axminster, runs to Plymouth, especially serving the south coast to the east of Exmouth, with important branches to Barnstaple and Ilfracombe and to Bude, and with a continuation into Cornwall. The Great Western Railway enters Devonshire at two points; near Burlescombe, running thence to Plymouth and into Cornwall, and near Venn Cross, for Barnstaple. The Cornish Riviera Express from Paddington, which slips a coach at Reading, and, passing south of Bristol, slips another at Exeter, performs the journey of 225 miles to Plymouth —the longest non-stop run of any train in England—in 7 minutes over 4 hours, which is an average speed of 55 miles an hour.

There are some famous bridges on the Devonshire roads and railways, of which the most remarkable are the Saltash Viaduct, 2240 feet long and 102 feet above high-water mark, built by Brunel across the Tamar; the old stone bridge of 16 arches over the Taw at Barnstaple, originally built in the thirteenth century, but since much altered and widened; the fifteenth century stone bridge of 24 arches over the Torridge at Bideford, also much changed from the days when it was only wide enough for a pack-horse, but always a valuable source of revenue to the town from the money that has, at various times, been left for its maintenance, and has been used to promote education, municipal improvements, charity and other objects; and the wooden bridge over the Teign at Teignmouth, one of the longest of its kind in England. Very different in character are the Lydford Bridge, whose single arch of stone spans the deep gorge of the river, close to the town; and the ancient stone clapper bridges, of which perhaps Post Bridge is the best known, already described in the chapter on antiquities.

## 24.   Administration and Divisions— Ancient and Modern.

In the days of our ancestors the Anglo-Saxons, Devonshire was governed much in the same way as it is governed now. That is to say, while the people had to obey the laws that were drawn up under the direction of the King, they had a great deal of what we now call self-government. Every little group of houses in Devonshire had its own "tun-moot" or village council, which made its own by-laws (from the Danish *by*, a town) and managed its own affairs. The large divisions of the county called Hundreds—groups of a hundred families—had their more important "hundred-moot"; while the general business of the whole shire was conducted by the "shire-moot," with its two chief officers, the "ealdorman," or earl, for military commander, and the "shire-reeve" for judicial president. The Devonshire shire-moot met twice in the year. These three assemblies may fairly be said to correspond to the Parish Councils, the District Councils, and the County Council of the present time. Our lord-lieutenant corresponds to the ealdorman of other days, and the present sheriff to the ancient shire-reeve.

The division called a Hundred may have been named, as already suggested, because it contained a hundred families. But the present Devonshire Hundreds, of which there are 32, vary a good deal in population. The Hundred of Black Torrington, for example, contains 38 parishes, and the Hundred of Hemyock only three.

The Parish is another ancient institution, and was originally "a township or cluster of houses, to which a single priest ministered, to whom its tithes and ecclesiastical dues were paid." Many of the 516 ecclesiastical parishes or parts of parishes situated wholly or partly within the Ancient Geographical County of Devon fairly correspond to the manors described in Domesday Book; but the whole country was not divided up into parishes until the reign of Edward III. The parishes, again, vary much in size and population. Thus, the parish of Lydford, which includes a large part of Dartmoor, and measures more than 50,000 acres, being the largest parish in England, contains 325 inhabited houses and a population of 2812. The parish of Haccombe, on the other hand, contains three inhabited houses and nine people.

Queen Elizabeth made the parishes areas of taxation, partly, at any rate, to provide funds for the relief of the poor. In modern times, with the idea of taking still better care of the poor, the parishes have been grouped together in Poor Law Unions, of which there are 20 in Devonshire, each provided with a workhouse, which was meant to be a place in which the able-bodied poor might find employment. Now, however, the workhouse is little more than a refuge for the destitute, the idle, and the incapable.

The local government of Saxon times was swept away by the feudal system of the Normans, which transferred the power of making and carrying out laws from the freemen to the lords of the various manors, and was only restored as recently as 1888 and 1894.

The affairs of each parish, since the latter date, have been managed by a Parish Council of from 5 to 15 men or women, elected by the parishioners. District Councils have charge of wide areas, and have larger powers. They are, in particular, the sanitary authorities, and are responsible for the water-supply. The County Council, whose very considerable powers extend to the whole shire, is a small parliament, which can levy rates and borrow money for public works. It manages lunatic asylums and reformatories, keeps roads and bridges in repair, controls the police in conjunction with the Quarter Sessions, appoints coroners and officers of health, and sees that the Acts relating to local government are carried out.

The Devonshire County Council consists of 103 members, of whom 77 are elected every three years by the ratepayers of the various electoral districts; while 26 are aldermen, elected or co-opted by the 77; thirteen of them in one triennial period, to serve for six years, and the other thirteen in the next period, to serve for the same length of time. The Council meets at Exeter, four times in the year. Plymouth, Devonport, and Exeter are called County Boroughs, and their corporations have the powers of a County Council. Ten other towns, Barnstaple, Bideford, Dartmouth, Great Torrington, Honiton, Oke-hampton, South Molton, Tiverton, Torquay, and Totnes, are called Municipal Boroughs and are governed by a mayor and corporation.

For the administration of justice the county, which is in the Western Circuit, has one court of Quarter

The Guildhall, Exeter

Sessions, the Assizes being held at Exeter; while Petty Sessions, presided over by local justices of the peace, are held weekly in 22 towns, to try cases and to punish those who have broken the law.

Ecclesiastical affairs are in the hands of the Bishop of Exeter, the archdeacons of Barnstaple, Exeter, and Totnes, together with numerous deans and other church officials, in addition to the parish clergy.

The County Council appoints a number of Education Committees, who have charge of all Government elementary and secondary schools throughout the county.

Devonshire is divided into eleven Constituencies, of which eight are Parliamentary Divisions, known as those of Honiton, Tiverton, South Molton, Barnstaple, Tavistock, Totnes, Torquay, and Ashburton, each of which returns one member. In addition to these Plymouth and Devonport each return two members and Exeter one, so that the county is represented altogether by thirteen Members of Parliament.

We may recall with pride the fact that, among the Members for Devon, have been some of the most distinguished men who have ever sat in Parliament. Thus, Sir Walter Ralegh sat for the county, Plymouth has been represented by Sir Francis Drake, Sir John Hawkyns, and Sir Humphrey Gilbert, Tavistock by John Pym and Lord John Russell, Barnstaple by Skippon and Lord Exmouth, Okehampton by William Pitt and Lord Rodney, Plympton by Lord Castlereagh and Sir Christopher Wren, Dartmouth by Lord Howe, and Tiverton by Lord Palmerston.

## 25. The Roll of Honour of the County.

Famous as our county is for its beautiful scenery, its wealth of prehistoric antiquities, and the abundance and variety of its wild life, it is still more renowned for its long Roll of Honour, for the many great and distinguished men who were born in it, or who have been more or less closely associated with it by residence within its borders. There can be little doubt that the foremost man in the whole history of Devon is Sir Francis Drake, the greatest of Elizabethan seamen, the first English circumnavigator of the globe, the terror of every Spanish ship, and the most conspicuous figure in the defeat of the Armada. Born near Tavistock, about 1540, of humble parentage, he early took to the sea, and was only seventeen when he and his kinsman Hawkyns, in the course of a trading-voyage to Guiana, were so ill-treated at a South American Spanish port that, for the rest of his life, Drake's chief aim seems to have been to avenge his injuries by plundering the towns, destroying the shipping, and capturing the treasure-ships of both Spain and Portugal.

Among his greatest exploits were his voyage round the world, between 1577 and 1580—after which, on his return, he was knighted by Queen Elizabeth on board his ship the *Golden Hind*; the ravaging of the West Indies in 1585 and 1586; the destruction in the harbour of Cadiz of ships and stores intended for the invasion of

England, by which he delayed for a whole year the sailing of the Armada; and the prominent part he took

Sir Francis Drake

in the defeat of the Armada itself, when he captured the flagship of Admiral Pedro de Valdez.

During the comparatively few years he spent on shore Drake constructed the still-existing leat or watercourse for bringing drinking-water into Plymouth, and he also represented that town as Member of Parliament.

In 1595 he and Hawkyns set out for the West Indies on what proved to be their last expedition. Misfortune dogged the fleet from the outset. Both commanders died at sea, Hawkyns off Porto Rico, late in 1595, and Drake off Porto Bello, early in 1596.

A greater man in some ways than even Drake himself was the gentle, noble, lovable, gallant Sir Walter Ralegh, a man who won renown in many fields, not only as soldier, sailor, and explorer, as courtier and administrator, but as historian and poet; whose whole life was crowded with adventure and romance, and who is one of the most picturesque figures in the entire range of English History.

Born in 1552, in a house that still stands at Hayes Barton, he was only 17 when he left Oxford to fight for the Huguenots; and from that time, except for brief intervals at Court, and even shorter periods of quiet enjoyment of his property in Ireland, or of his home at Sherborne, or when he was a prisoner in the Tower, the rest of his life was spent in action; now fighting the rebel Desmonds in Ireland, now harrying the ships and towns of Spain and Portugal, now helping in the attack on the Armada, now engaged with his half-brother Sir Humphrey Gilbert in perilous and fruitless exploration in the far north of America, now attempting to colonise Virginia—an enterprise whose sole result was the introduction to this country of tobacco and potatoes —and now sailing up the Orinoco in the vain quest of the fabled golden city of Manoa.

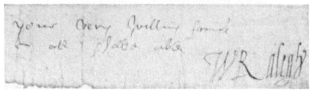

Sir Walter Ralegh and his signature

Elizabeth, whose favour he won by the sacrifice of his cloak, and lost again for a time owing to her jealousy of his passion for one of her Maids of Honour, when he had to spend four years in the Tower, knighted him, gave him vast estates in Ireland, made him captain of the guard, Governor of Jersey, Lord Warden of the Stannaries and Vice-Admiral of Devon and Cornwall. Like Drake, he sat in Parliament; and it was while still in favour with the Queen that he was elected Member for his county.

On the accession of James I, however, Ralegh was charged with joining in the plot on behalf of Arabella Stuart, and was again sent to the Tower. During his long imprisonment there he wrote his most famous work, the *History of the World*, whose learned, eloquent, and philosophic pages proved that his skill was no less with his pen than with his sword. His stirring description of the last fight of the *Revenge* inspired Tennyson's noble ballad.

Released from prison by James in order that he might once more sail up the Orinoco in search of the mythical treasure-land ruled over by El Dorado, he came back from that most disastrous expedition a broken man. Again committed to the Tower at the instigation of the Spanish Ambassador, he was soon afterwards beheaded on the old charge of treason, dying as he had lived, dignified, noble, and fearless to the last.

Two other heroic figures of the Elizabethan age, worthy to be ranked in the same company with Drake, are his gallant comrade Hawkyns, who was born at Plymouth in 1532, and Grenville the indomitable, the hero of that last fight of the *Revenge*.

Several other men who were born in our county have distinguished themselves as explorers, or by having helped, by peaceable means, to found our over-seas empire. Such were Davis, the arctic navigator, who was born near Dartmouth about 1550, who left his name in Davis's Straits, and who wrote *The Seaman's Secrets* and other works ; Sir Humphrey Gilbert, born in 1539 at Dartmouth, distinguished as a soldier in the Irish wars of Elizabeth's reign, but still more as having taken possession of Newfoundland, thus establishing the first British colony ; Gate, who with Somers colonised Bermuda in 1611 ; and Wills, who perished in 1861 with Burke in crossing Australia.

Devonshire has been the native land of many soldiers. Two of the most distinguished, both of whom strongly influenced their country's destiny, and were made dukes as a reward for their services, were Monck and Marlborough. George Monck, born near Torrington in 1608, distinguished himself both by land and sea. He twice defeated the great Dutch admiral Van Tromp ; and although severely beaten by de Ruyter he afterwards gained a great victory over him off the North Foreland. At first a Royalist, he joined the parliamentary army after his capture by Fairfax (followed by two years in the Tower) and Cromwell made him governor of Scotland. On the death of the Protector he marched to London, and was the chief instrument in the Restoration of Charles II, who made him Duke of Albemarle.

John Churchill, better known as the Duke of Marlborough, born at Ashe House in 1650, was not only the greatest general of his time, but one of the ablest military commanders the world has ever seen. His most memorable successes were the four great battles of Blenheim, Ramillies, Oudenarde, and Malplaquet, in which he defeated the long-victorious armies of Louis XIV, then the most powerful monarch on the continent. By this series of victories, followed by the Treaty of Utrecht in 1713, the peace of Europe was secured for thirty years.

Many great Devonshire men, including some of the earliest who became distinguished, were churchmen or divines, not a few of whom are also famous as authors. Such, for example, was Winfrid, otherwise Saint Boniface, and known as the Apostle of Germany, who, born probably at Crediton in 680, began his career as a Benedictine monk at Exeter, and after spending many years in converting the wild German tribes to Christianity, was appointed archbishop of Mainz, and was afterwards murdered by the Frisians in 755. Such were Leofric, the first bishop of Exeter, Warelwast the builder of the Norman cathedral, Quivil the designer of the magnificent fabric that replaced it, Stapledon and Grandisson his able successors, Reynolds the leading Puritan divine at the Hampton Court Conference of 1604, whose proposal of a new translation of the Bible led to the Authorised Version of 1611, Trelawney, one of the Seven Bishops whose trial and acquittal formed one of the most memorable events of the reign of James II, Jewel, Bishop of Salisbury, one of the fathers of English Protestantism, the "Judicious" Hooker, author of the *Laws of Ecclesiastical Polity*, Barclay, who translated Brant's satiric allegory under the title *The Shyp of Folys*, Prince, author of the *Worthies of Devon*, Dean Buckland the famous geologist, author of *Reliquiae Diluvianae*, who died in 1856, and Charles Kingsley, born at

Charles Kingsley

Holne in 1819, distinguished as an able and eloquent preacher and as a strenuous worker for the good of mankind, as poet, novelist, and naturalist, author of many books, and especially of *Westward Ho!* and the *Water Babies*, and of the words of many beautiful songs, such as the

*Three Fishers.* Not a native of the county, but Bishop of Exeter in 1551, was Miles Coverdale, whose translation of the Bible appeared in 1535. To him many of the finest phrases in our Authorised Version of 1611 are directly due.

The most distinguished of the many Devonshire men of letters is Coleridge, poet and dreamer, philosopher and critic, who was born at Ottery St Mary in 1772. That, however, was his sole connection with the county. It was chiefly during his three years' residence at Nether Stowey, in Somerset, that the finest of his few master-pieces, especially the *Ancient Mariner* and part of *Christabel*, were written. Amongst other authors who were born in Devon may be named Gay, writer of plays, fables, and songs, among them the *Beggars' Opera* and *Black-eyed Susan*; Ford the dramatist; William Browne, the author of *Britannia's Pastorals*; Kitto, the deaf compiler of Biblical literature; Merivale the Roman historian; Rowe and Risdon, each of whom wrote books on the county; and Froude the historian, author of many books, and especially of the *History of England from the Fall of Wolsey to the Defeat of the Spanish Armada.*

Herrick was not Devonshire born, but it was while he was vicar of Dean Prior, between 1647 and 1674, that he wrote the *Hesperides*, among which are some of the best lyrics in the language. Dryden, again, was a frequent visitor at Lord Clifford's seat at Ugbrooke, and there is a tradition that he there finished his translation of Virgil. It was at Lynton that Shelley wrote part of *Queen Mab*. Keats finished *Endymion* at Teignmouth. Tennyson was often a guest of Froude at Salcombe, and it is said that he had Salcombe Bar in mind when he wrote his last verses, *Crossing the Bar.*

Distinguished in other ways may be mentioned Blundell, the Tiverton cloth-merchant, who, dying in 1601, left money for the establishment of Blundell's School; Bodley, born at Exeter in 1545, founder of the Bodleian Library at Oxford; John Baring, founder of

Blundell's School, Tiverton

the great banking-house of Baring Brothers; Babbage, the inventor of the calculating machine; Bidder, the "Calculating Boy," son of a stone-mason of Moreton Hampstead; Cookworthy, the originator of Plymouth china; and Newcomen, a Dartmouth ironmonger, whose improvement on the atmospheric steam-engine of Savery, also a Devonshire man, was used early in the eighteenth century for pumping water out of mines.

Devonshire has been specially remarkable for its artists, of whom the most distinguished were Sir Joshua Reynolds, the great portrait-painter, born at Plympton in 1723; Cosway, who painted exquisite miniatures; Samuel Prout, the famous architectural painter, and

Samuel Taylor Coleridge

Skinner Prout his nephew; Eastlake, the great painter of figures, and the author of books on art; and Hilliard the goldsmith of Queen Elizabeth.

Two very famous Devonshire houses are those of Courtenay and Carew. There is said to be hardly a parish in all Devon in which a Courtenay did not hold land. Courtenays followed the King to many wars. One tilted with Francis I at the Field of the Cloth of Gold. Three were at Navarete with the Black Prince. Three died during the Wars of the Roses, either in battle or on the scaffold. Of the house of Carew, one was at Cressy and another at Agincourt. One was knighted on the field of Bosworth, one was at Flodden, and one, while fighting the French, was blown up with the *Mary Rose*.

## 26. THE CHIEF TOWNS AND VILLAGES OF DEVONSHIRE.

(The figures in brackets after each name give the population of the parish in 1901, from the official returns, and those at the end of each paragraph are references to the pages in the text.)

**Appledore** (2625). A small sea-port at the mouth of the Torridge, wrongly supposed, through confusion with an Appledore in Kent, to have been the landing-place of Hubba the Dane. (pp. 27, 61, 130, 131.)

**Ashburton** (2628). A market-town on the Yeo, eight miles south-west of Newton Abbot, one of the Stannary Towns, with some manufacture of cloth. A good centre for Dartmoor, and with a fine church and other old buildings. Near it are Holne Chase and the Buckland Woods, with very beautiful scenery. (pp. 46, 112, 118, 120, 151, 173, 212.)

**Axminster** (2906). Close to the border of Dorset, high above the Axe. Interesting for the history of its church, founded in 755, and endowed by Athelstan after his victory over the Danes. The manufacture of Axminster carpets was discontinued here in 1835. (pp. 114, 150, 153, 168, 170, 199, 206.)

**Axmouth** (643). A pretty village in a combe in rugged chalk cliffs, near the mouth of the Axe. The coast here has been much altered by landslips. (pp. 66, 199.)

**Bampton** (1657). An old market-town, near the border of Somerset, with a very large annual fair, especially for the sale of sheep and Exmoor ponies. Has also large limestone quarries. (p. 191.)

**Barnstaple** (11,999). The chief town of North Devon, nine miles from the mouth of the Taw, where the river widens into a tidal estuary. It formerly had much trade with America, but is now noted only for its pottery, called Barum ware. The river is spanned by a famous stone bridge of 16 arches, dating from the thirteenth century. (pp. 82, 101, 112, 114, 130, 131, 139, 149, 173, 192, 206, 207, 210, 212.)

**Beer** (1118). A fishing-village at the foot of a narrow, deep valley near the Dorset border, noted for lace-making, and for its very extensive subterranean quarries of fine building-stone. (pp. 68, 112, 114, 123, 168, 200.)

**Bere Alston.** A village eight miles north of Plymouth, close to the border of Cornwall, was formerly noted for its rich silver mine, flooded by the Tamar in 1860. (p. 121.)

**Berry Pomeroy** (423). A village in the valley of the Dart, near Totnes, famous for its ruined castle, the most picturesque ruin in Devonshire. (pp. 100, 176, 188.)

**Bideford** (8754) is a market-town and river-port near the mouth of the Torridge, here crossed by a fine bridge, built in the fifteenth century in place of the dangerous ford which gave its name to the town. It was a very important place in Armada days, and formerly had great trade with Newfoundland and other American colonies. (pp. 114, 123, 130, 131, 194, 204, 207, 210.)

**Bovey Tracy** (2693), six miles north-west of Newton Abbot, is noted for beds of clay and lignite, and for its potteries. (pp. 25, 40, 114, 122, 123.)

**Brixham** (8092), a sea-port with a good harbour, a market-town, and a very important fishing-station, with many trawlers, stands on Berry Head, at the south end of Torbay. Here William of Orange landed in 1688.   (pp. 73, 118, 128, 130, 132, 150, 154, 197.)

**Buckfastleigh** (2781) is a small town in the Dart valley, with woollen factories. Buckfast Abbey, a Saxon foundation, was restored and reinhabited by French Benedictine monks in 1882.   (pp. 112, 171, 184, 200.)

**Buckland Abbey**, seven miles north of Plymouth, was in part converted into a dwelling-house by Sir Richard Grenville, and this was afterwards altered by Sir Francis Drake, of whom interesting relics are here preserved.   (pp. 54, 145, 166, 184.)

**Budleigh Salterton** (1883). A small port and favourite watering-place, beautifully situated five miles east of the mouth of the Exe.   (pp. 69, 129, 199.)

**Chagford** (1397). A small market-town, high above the Teign valley, on the borders of Dartmoor, forming a good centre for tourists, naturalists, and archaeologists. There are many Bronze Age antiquities in the neighbourhood.   (pp. 25, 120, 191, 204.)

**Chudleigh** (1820), seven miles inland from Dawlish, contains the ruins of the palace of the Bishop of Exeter, built in 1080. Ugbrooke, often visited by Dryden, is a mile away.   (pp. 123, 199, 200, 221.)

**Chumleigh** (1158), is a village on high ground above the valley of the Taw, chiefly interesting for the history of the Seven Prebends of its church.   (pp. 172, 196, 197.)

**Clovelly** (621). A small but extraordinarily picturesque fishing-village, consisting of one cobble-paved street, running steeply up a narrow ravine through a densely-wooded hill-side.

Clovelly

Near it is the Hobby Drive.  There is a fine camp on the hill above.  (pp. 62, 63, 85, 161, 176.)

**Colyton** (1943) is a small market-town, beautifully situated in the Coly valley, near the border of Dorset, with a fine church. (pp. 114, 122, 150, 168, 191.)

**Combe Martin** (1521), a village on the coast six miles east of Ilfracombe, in a fertile valley, was formerly noted for its very rich silver mine; now for market-gardening.  (pp. 15, 121, 171, 172.)

**Countisbury** (279) is a little village on the west side of the Foreland, close to the Somerset border.

**Crediton** (3974), a market-town with boot and shoe, and cider factories, stands above the valley of the Creedy, eight miles north-west of Exeter, whither the see of the Bishopric was, for greater safety, moved from here by Leofric, in 1050.  The very fine church, of unusual length, contains many monuments.  (pp. 112, 118, 168, 172, 200, 219.)

**Cullompton** (2919) is a market-town 12 miles north-east of Exeter, on the road from Bristol.  The manor belonged to Buckland Abbey.  The Walronds is a fine Elizabethan mansion. (pp. 112, 118, 172, 173.)

**Dartington** (478), so-named when the tidal estuary of the Dart ran close to it, is now a suburb of Totnes.  Dartington Hall is a very fine Elizabethan house.  (pp. 129, 194.)

**Dartmouth** (6579), a market-town, and favourite resort of yachtsmen, and formerly a port of great importance, at the narrow entrance of the Dart estuary, is a place of exceptional beauty and of great historic interest, built in terraces on a steep, wooded hill. In the old town along the quay and in the Butter-Walk are fine old Elizabethan houses.  St Saviour's, one of its four churches,

Dartmouth, from Warfleet

dates from 1372, and has a splendid rood-screen and a very fine pulpit. The land-locked harbour was guarded by two castles. On a hill above the town is the great white building of the naval college, which has superseded the old training-ship *Britannia*. (pp. 23, 74, 75, 129, 130, 132, 140, 147, 172, 175, 176, 191, 200, 210, 212, 218, 222.)

**Dawlish** (4287) is a charming and highly popular watering-place with fine sands and beautiful red cliffs, in a sheltered combe, south of the estuary of the Exe. A pretty pleasure garden called the Lawn, with a stream through it, divides the new town from the old. (pp. 69, 80, 128.)

**Devonport** (70,437) is a parliamentary, municipal, and county borough, on high ground above the estuary of the Tamar, two miles west-north-west of Plymouth, one of the chief naval arsenals in Britain, with government establishments—dockyards, barracks, magazines, etc.—stretching nearly four miles along the Hamoaze, a great anchorage for men-of-war. (pp. 79, 118, 210, 212.)

**Drewsteignton** (673), a large village near the Teign, not far from which is the Spinster's Rock, the only cromlech in Devonshire. (p. 154.)

**Exeter** (47,185), the capital of Devonshire, and long regarded as the Key of the West of England, is a picturesque old city, standing on high ground above the Exe, which passes through the town. It is a port, with a large basin connected with the estuary of the Exe by a canal. A municipal, county, and parliamentary borough, it has factories of agricultural implements and gloves, and there are large nurseries round it. Of its castle of Rougemont, built by the Conqueror in 1067, little now remains. But its magnificent cathedral, which contains many most beautiful and interesting features, is the finest example of the Decorated style of architecture in England. Other interesting or important buildings are the Deanery, the House of the Abbots of Buckfast, Mol's Coffee House, the Guildhall, and the Albert Memorial Museum.

The history of the town is of the highest interest, and is linked with every event of importance connected with the county. It has been besieged in turn by Danes and Normans and Saxons, by King Stephen, by the army of Perkin Warbeck, and the rebels of the "Commotion," by the Yorkists, and by Royalists and Parliamentarians. Many distinguished Bishops have held the see; and noteworthy names of those who have been born in the city are those of the "Judicious" Hooker, Sir Thomas Bodley, and the Princess Henrietta, daughter of Charles I, and afterwards Duchess of Orleans. (pp. 22, 101, 112, 118, 132, 137, 138, 139, 140, 141, 142, 146, 147, 148, 149, 150, 160, 163, 164, 168, 170, 171, 175, 176, 178, 180, 181, 182, 187, 195, 196, 197, 200, 203, 204, 205, 210, 212, 219, 221, 222.)

**Exminster** (2550), a village on the west bank of the Exe, where is a large asylum.

**Exmouth** (10,485), once a sea-port, is a rapidly growing and very popular watering-place, with docks and brick-works, at the entrance of the estuary of the Exe, here narrowed to a swift current by the sand-bank called the Warren. (pp. 22, 82, 96, 129, 132, 191, 206.)

**Haccombe** (9), the smallest parish in England, contains the residence, not always occupied, of the Carew family, and three other houses. In the tiny church, the rector of which is an "arch-priest," are many memorials. (pp. 171, 176, 209.)

**Hartland** (1634), a small town in a very large but thinly-inhabited parish, stands on the west side of Hartland Point. In the church, whose lofty tower serves as a steering-mark for ships in the Bristol Channel, is one of the longest rood-screens in Devon, and also a good Norman font and door. (pp. 64, 150, 170, 172, 173, 175, 184.)

Holbeton (850) stands on the Erme estuary, ten miles from Plymouth. Its fine church contains a magnificent rood-screen and a Norman font. (pp. 175, 196.)

Holcombe Rogus (607), on the Somerset border, southwest of Wellington, contains a church in which are many monuments of the Bluett family, who formerly owned Holcombe Rogus Court, the finest Tudor mansion in the county. (p. 195.)

Holne (273), a small village in the beautifully-wooded valley of the Upper Dart, was the birth-place of Charles Kingsley. The church has a fine screen and pulpit. (pp. 23, 160, 175, 220.)

Holsworthy (1371), is an important market near the border of Cornwall, ten miles inland from Bude. The very ancient horse-fair of St Peter is held here in July.

Honiton (3271), a municipal borough on the London and Exeter road, 16 miles from the latter town, gives its name to the lace which was first made here by Flemish refugees. St Margaret's Hospital for Lepers has been converted into almshouses. Four miles away is Hembury fort, one of the finest camps in Devon. (pp. 112, 138, 161, 175, 210, 212.)

Ilfracombe (8557), a small sea-port and very popular watering-place on the north coast, having a land-locked harbour sheltered by the Capstone Hill, and with the Chapel of St Nicholas, now a lighthouse, at the entrance of it, is celebrated for the exceptional mildness of its climate. (pp. 58, 94, 130, 131, 150, 151, 170, 206.)

Instow (634), a small but very ancient port, at the point where the Taw and the Torridge meet, has weekly communication with Lundy. (pp. 27, 61.)

Kenton (1612), a very picturesque village, inland from Starcross, with a fine church of red sandstone, whose rood-screen, partly Flemish, is one of the best in England, and its oaken pulpit perhaps the finest in Devon. (pp. 172, 174, 175.)

**Kingsbridge** (3025 with **Dodbrooke**) is a small but important market-town at the head of the Kingsbridge estuary, which is really a tidal creek without a river, in the extreme south of the county. It is one of the chief places in the fertile district called the South Hams. (pp. 16, 50, 54, 118, 122, 193.)

**Kingswear** (841) is a picturesque village opposite Dartmouth, which is reached from it by a steam-ferry. Near the old

Cherry Bridge, near Lynmouth

castle, now modernised, but said to date from John's reign, are the remains of a guard-house from which a chain was stretched across the river to Dartmouth Castle, to guard the estuary. (pp. 74, 191.)

**Lydford** (2812), a small village in the largest parish in England, including a great part of Dartmoor, was once second in importance to Exeter, a Stannary Town, and the seat of the Stannary prison. There is a ruined Norman castle. Lydford

Gorge, spanned by a single-arched stone bridge, is one of the most beautiful spots in Devon. (pp. 14, 120, 138, 139, 190, 204, 207, 209.)

**Lynton** (1641) and **Lynmouth** (402) are two villages in the parish of Lynton, on the north coast, the latter on the shore, and the former 450 feet above it, famous for their very beautiful

Lynmouth Harbour

scenery, especially along the river Lyn—where one of the finest spots is at the Watersmeet—and in the wild ravine called the Valley of Rocks. (pp. 16, 57, 82, 127, 150, 221.)

**Modbury** (1242) is a small market-town 12 miles south-east of Plymouth, once the principal residence of the Champernownes, who made it famous as a musical centre in Tudor times.

**Moreton Hampstead** (1541), a picturesque little town on the eastern border of Dartmoor, with an important cattle-market. (pp. 151, 222.)

**Mortehoe** (788), a small but growing watering-place near Ilfracombe, with an interesting church, and not far from Woollacombe Sands and the dangerous headland of Morte Point. (pp. 60, 173.)

**Newton Abbot** (16,951), in very beautiful country six miles from Torquay, has large markets for cattle and for dairy-produce, and wharves on the Teign for trade in timber and coal. The parish church has fine screens and many monuments. Both Charles I and the Prince of Orange were entertained here at Forde House. (pp. 26, 161, 195, 197, 203, 205.)

**Okehampton** (2569), on the north-west edge of Dartmoor, has large markets for cattle and agricultural produce. In the neighbourhood are the very picturesque ruins of a Norman castle and other attractions both for antiquarians and naturalists. (pp. 140, 151, 188, 210.)

**Ottery St Mary** (3495), a market-town south-east of Exeter, in the beautiful valley of the Otter, is famous for its noble church, the finest in Devonshire, and containing many very interesting and beautiful features, and also as being the birth-place of the poet Coleridge. (pp. 170, 171, 173, 176, 178, 195, 200, 221.)

**Paignton** (8385), a rapidly-growing watering-place on Torbay, with a fine situation, a bracing climate, and good sands (pp. 73, 175, 176.)

**Plymouth** (107,636), very finely situated at the mouth of the river Plym, at the head of Plymouth Sound, a parliamentary, municipal, and county borough, the chief seat of trade, commerce, and manufactures in Devonshire, is one of the most famous seaports in the kingdom. Described in Domesday Book as

Sutton, and occasionally known as Plymouth as early as the fourteenth century, it did not definitely receive its present name until the reign of Henry VI.

Its spacious docks, Millbay, the graving-dock and the floating basin, can accommodate the largest merchant-ships. In Sutton Pool and the Catwater, in the Hamoaze—the estuary of the Tamar, the Lynher and the Tavy—and at the head of the Sound,

Ogwell Mill, near Newton Abbot

in the shelter of the breakwater, a very large number of vessels find safe and convenient anchorage. At its numerous quays, connected with the Great Western Railway, to which company the docks belong, are landed passengers and mails from the United States, from Australia and New Zealand, from the West Coast of Africa and the Cape, from India and the East, as well as merchandise from all parts of the world, especially from France. Next to Newlyn, it is the most important fishing-station on the south coast of England. Further details will be found in the chapters on shipping and fisheries. Two of its most remarkable monuments, both on the Hoe, are a copy of Boehm's fine statue of Drake, and part of the old Eddystone lighthouse, re-erected as a memorial to Smeaton.

Plymouth has had a stirring history. In mediaeval times, especially in the fourteenth and fifteenth centuries, it suffered much from the attacks of the French, who, in 1403, under du Chastel, are said to have burnt 600 houses. Its most important periods are those connected with the defeat of the Armada, with the Civil War, and with the French war that ended with the battle of Waterloo. It was in the Catwater that the English fleet lay at anchor, while Drake and his fellow-captains waited on the Hoe, the famous ridge between Millbay and Sutton Pool, until the Spanish ships had passed. More or less closely blockaded from 1642 to 1646 by the Royal forces, and many times desperately assailed, Plymouth was the one town in the whole west of England that was never lost to the Parliament. In the Napoleonic war the town was the scene of great activity, fitting out many naval expeditions against the French, and receiving many captured ships.

Many famous names are associated with the town. It was from here that the Black Prince set out for France and the victory of Agincourt. Here, in 1470, landed the Duke of Clarence, in the hope of enlisting recruits for the Lancastrian army. Here, too, came Margaret herself, with Prince Edward, just before the final overthrow at Tewkesbury. It was at Plymouth that the Princess of Aragon landed, on her way to marry Prince Arthur. From Plymouth sailed Drake and Hawkyns on their filibustering expeditions, and to this port they came back loaded with Spanish gold. Here, too, came Drake, after his voyage round the world. From here Sir Humphrey Gilbert set out on his last voyage, and from here sailed Captain Cook. In the streets and on the quays of Plymouth Benbow and Rodney, Howe and Jervis, Collingwood and Nelson, were, in their time, familiar figures. (pp. 48, 78, 79, 82, 83, 86, 101, 114, 123, 128, 130, 131, 132, 134, 136, 140, 142, 144, 146, 147, 148, 151, 153, 159, 160, 199, 204, 205, 206, 207, 210, 212, 215, 217, 222.)

**Plympton** (4954), a parish north of Plymouth, comprising two separate villages which grew up round the castle and the priory. Until the fifteenth century the Prior of Plympton controlled the affairs of Plymouth. (pp. 108, 114, 172, 183, 190, 212, 223.)

**Plymstock** (3195), a parish to the east of Plymouth, with large quarries, and with extensive fortifications for the protection of the harbour.

**Powderham** (233), a village on the west side of the estuary of the Exe, where, in a very beautiful park, stands Powderham Castle, chief seat of the Courtenays, Earls of Devon. (pp. 187, 192.)

**Princetown,** in the western part of Dartmoor, is the site of a famous convict prison, originally built, in 1809, for the reception of French prisoners of war. The convicts have brought much land into cultivation, and there are also large granite quarries in the neighbourhood. (pp. 96, 151, 155, 205.)

**Salcombe** (1710), a small port at the mouth of the Kingsbridge estuary, with an exceptionally mild climate, and with other attractions as a watering-place. (pp. 82, 132, 149, 191, 222.)

**Seaton** (1325), a pleasant watering-place near the mouth of the Axe, in the chalk cliffs, close to Dorset. (pp. 66, 80, 114.)

**Shute** (461), a scattered parish, containing the former seat of the De la Poles, has many monuments to them in its church.

**Sidmouth** (4201), a fashionable watering-place, very pleasantly situated at the mouth of the Sid, between Exmouth and the border of Dorset. Its equable climate is perhaps its chief attraction, but it was an important harbour before its sheltering cliffs were destroyed by landslips. Queen Victoria spent some years of her childhood here. (pp. 68, 80, 130, 163.)

**South Molton** (2848), a very ancient market-town in the south of Exmoor, has corn-mills and a very fine church-tower. (pp. 172, 212.)

Shute Manor House

**Tavistock** (4728), close to the border of Cornwall, with the ruins of a great abbey, round which the town grew up, was formerly very famous as a mining-centre, and was one of the Stannary Towns. Sir Francis Drake was born here, and the statue of him by Boehm, of which there is a copy on Plymouth Hoe, is one of that sculptor's finest works. (pp. 112, 118, 120, 122, 138, 151, 171, 183, 197, 204, 205, 212, 213.)

**Teignmouth** (7366), an ancient sea-port, a modern and very popular watering-place and a market-town at the mouth of the Teign, has a good harbour, sheltered by the Den, once a mere sandbank, but now a promenade and pleasure-garden. The wooden bridge over the river is one of the longest in England. (pp. 26, 69, 71, 129, 132, 140, 150, 207, 221.)

**Bickleigh** (10,382), an old market-town where the Loman joins the Exe—hence the name, Two-ford-town—was formerly

Bickleigh Bridge

noted for its woollen trade, but now for its lace-factory. The church contains many interesting monuments. (pp. 111, 114, 146, 163, 166, 172, 176, 190, 200, 202, 205, 210, 212, 222.)

**Topsham** (2790), once a famous port, is now a market-town and fishing station on the estuary of the Exe. (pp. 22, 130, 132.)

**Torcross**, a small fishing-village and watering-place at the south end of the Slapton Sands. The bay being very exposed, the fishermen train Newfoundland dogs to swim out to boats in rough weather, and take the " painter " ashore. (p. 129.)

Torquay (33,625), one of the best-known towns in Devon, is a large and fashionable watering-place, very celebrated for its mild and equable climate, standing on the south slopes of the northern headland of Torbay. From the well-sheltered little harbour the town rises in a semicircle, so protected from rough winds that palms, myrtles, aloes, agaves and other sub-tropical trees flourish here freely in the open air. Near the town is Kent's Cavern, in which have been discovered many most interesting remains of extinct animals and of pre-historic man. (pp. 11, 73, 94, 95, 123, 127, 128, 132, 151, 153, 188, 210, 212.)

Torrington (3241), a market-town on the Torridge, south of Bideford, with important fairs and cattle-shows, and with factories of gloves. The storming of Torrington by Fairfax in February, 1646, was the death-blow to the cause of King Charles, and practically ended the Civil War in Devonshire. (pp. 118, 148, 149, 191, 210, 218.)

Totnes (4035), one of the oldest municipal boroughs in England, at the head of the navigable portion of the river Dart, is one of the chief market-towns of the South Hams, with a ruined castle and other remains of fortification, some picturesque old houses, a fine church with a specially good stone rood-screen, and a granite obelisk in memory of the Australian explorer Wills, who was born here. (pp. 23, 112, 139, 175, 188, 191, 194, 210, 212.)

Westward Ho! a watering-place on Northam Burrows on the shore of Barnstaple Bay, with a good climate, and with many attractions for the marine zoologist, was named in honour of Kingsley's great romance. (pp. 62.)

Widecombe-in-the-Moor (657), a village in the centre of Dartmoor, with an annual fair for the sale of sheep and ponies, with many very interesting prehistoric remains in the neighbourhood, and a very fine church tower. (pp. 167, 173.)

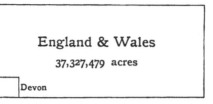

Fig. 1. The Area of the Ancient Geographical County of
Devon (1,667,154 acres), compared with that of England
and Wales

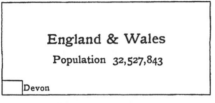

Fig. 2. The Population of Devon (661,314) compared with
that of England and Wales (in 1901)

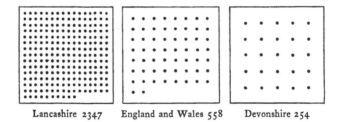

Fig. 3. Comparative Density of Population to Square
Mile in 1901

*(Each square represents a square mile)*

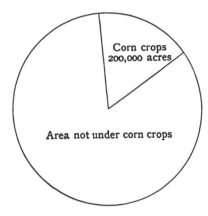

Fig. 4.   Proportionate Area under Corn Crops in
Devon in 1908

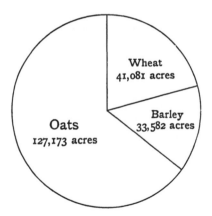

Fig. 5.   Proportionate Area of chief Cereals
in Devon in 1908

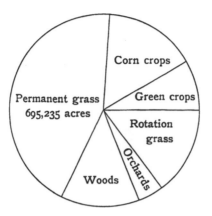

Fig. 6. Proportion of Perennial Pasture to other
Areas in Devon in 1908

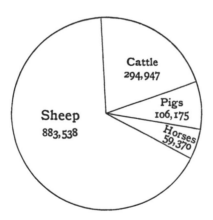

Fig. 7. Proportionate numbers of chief Live-stock
in Devon in 1908

Milton Keynes UK
Ingram Content Group UK Ltd.
UKHW041520181024
449640UK00009B/97

9 781107 690752